As someone who has been both excited by the rapid expansion of the scientific knowledge base in areas such as neuroscience, physics, biology, developmental psychology, systems thinking and consciousness, but also overwhelmed at the explosion of research and publications, this book is a godsend. It integrates much cross disciplinary material to address critical human developmental and learning processes. As an educator myself, I am already applying my new understandings in my teaching and training.

> — David Lukoff, PhD, Professor Emeritus of Psychology at Sofia University in Palo Alto, California; author or coauthor of 80 articles and chapters on spiritual issues in mental health services

* * *

Mindleap, by Jim Brown, PhD, reveals how the innovative methods of neuroscience can be applied to create a system of education that is better able to meet the challenges of our accelerating world. Essential to this endeavor, Brown emphasizes, is understanding the value of enhancing self-actualization of learners within the context of relationship. He invites not only psychologists, educators, graduate and post-graduate students to explore this new model of thinking, but it also encourages other seekers and creators to join together in shaping a culture of life-long learning.

His brilliant offerings are applicable to every day perceptual experiences, as well as to the development of new paradigms in education. He is a generous writer, passionate about translating complex systems of thought into a relatable scientific-poetic style. After constructing a solid foundation of current information and theory from brain research, he respectfully creates a covenant with his readers by translating the vast array of technical language into understandable metaphors and concrete life examples.

> — Susan S. Scott, PhD, Jungian Analytical Psychology practitioner, Author of *Healing with Nature* and eight published essays.

* * *

Mindleap is a breath of fresh air. Jim Brown has introduced a radically different approach to education, asserting that the current system is overly linear and reductionistic. He proposes an educational system based on current evidence from the neurosciences and insights from chaos and complexity theory. The educational reform he advocates would not only foster creativity but would help students thrive in a world that is neither linear nor reductionistic in nature, a world that is becoming more complex with each decade. Bravo to this author for his insight and prescription!

— Stanley Krippner, PhD, co-author, *Personal Mythology*, co-editor, *Varieties of Anomalous Experience*

MINDLEAP

A FRESH VIEW OF EDUCATION EMPOWERED BY NEUROSCIENCE AND SYSTEMS THINKING

by

JIM BROWN, PhD

FOREWORD BY
DON HANLON JOHNSON, PhD

PSYCHOSYNTHESIS PRESS
MOUNT SHASTA, CA

PSYCHOSYNTHESIS PRESS
Exploring the Deeper Possibilities of Human Life

MINDLEAP

A FRESH VIEW OF EDUCATION EMPOWERED BY NEUROSCIENCE AND SYSTEMS THINKING

by Jim Brown, PhD

This book, which not only advocates refinements in the paradigm of education but also shows that mind, body/brain, and relationship/environment are explicitly intertwined, is by its very existence a concrete illustration of that principle. My mind and body/brain could not have brought it into existence without that third element: relationship/environment. Without the relationship with my life-partner Molly and the environment issuing from our relationship, this book would have been impossible for me to create.

I therefore dedicate *Mindleap* to Molly Young Brown, whose spirit, skills and unflagging dedication to its completion enabled it to exist.

Table of Contents

Foreword   Reflections on *Mindleap*   1
                by Don Hanlon Johnson, PhD

Introduction Notes from the Frontier   3

Chapter 1   Applying Systems Thinking to Neuroscience   11
                and Education

Chapter 2   A Bit of History About Systems Thinking   17

Chapter 3   The Next Level of Systems Thinking—   27
                Complex Dynamical Systems

Chapter 4   Deepening Systems Understanding   47

Chapter 5   Expanding the View   59

Chapter 6   The Embodied Brain and Emergent Mind   75

Chapter 7   A Framework for Optimal Human Learning   89
                Throughout Lifelong Brain Development

Coda   Systems Perspectives on COVID-19   113
                and Beyond

Appendix A   Discussion of the Initial Five Phases   129
                Diagrammed in the Chapter 7 Table
                Optimal Human Learning and Development

Appendix B   Reflections on an Above Average   140
                Public School Education by Marika L. Foltz

                Bibliography   151

                Index   155

                Acknowledgements   161

                About the Author   163

Foreword

Reflections on *Mindleap*

by Don Hanlon Johnson, PhD

As the world darkens and my mood reflects a despair in the many destructive forces storming about us, this book gave me some cheer. It returned me to the beginnings of my now very long adult life of teaching, constructing institutions of learning, and healing. At the time of those beginnings, many of us were brought together from different realms of expertise sharing a realization that "learning" cannot occur successfully unless those in charge address serious questions about how people actually learn. If all the efforts of educators are focused on content (learning outcomes, skills to be achieved, materials to be memorized) and not on what is happening in the learner (a particular person's residues of history impacting their orientation to learn, family tragedies, illness, abuse and trauma, feelings of inadequacy) there is very little return on all the efforts expended on the process. Which is what has happened too frequently in the decades that have passed since the fertile time when so many innovations were occurring.

We also began our careers under the revelatory impacts of humanistic psychology, psychedelics, and transformative bodyworks which enormously expanded our understandings of what "reality" included as the horizon of any educational venture.

In light of that forgetfulness of what we thought was so obvious half a century ago, and how that forgetfulness has contributed to the thoughtless and conflictual period we are now confronted with, it is auspicious that this old veteran of a lifetime of raising these questions of how each of us actually learns most effectively has returned with new material to add to the analysis drawn from fresh insights afforded by advances in the neurosciences and systems thinking. These two communities of inquiry have added specificity to the more generalized notion of the human potential which motivated us half a century ago. In doing so, they both add heft to arguments in favor of cultivating the experiential, psychedelic, and energetic foundations of learning and suggest more precise strategies for so doing. A desperately needed nudge forward for those in control of designing the various institutions of learning.

I write these reflections in the midst of having spent the last 8 weeks with my new granddaughter, beginning with being present at her birth. I have spent hours with her as she awakens to this amazing world, wiggling, sounding, struggling, but most profoundly gazing…into my eyes, into the eyes of each of our family, and slowly more and more into the world about her. And with no agenda, just the openness like that of the clear sky. The very big question: how do we manage to protect and nurture this openness to the vastness of the real, instead of trying to squeeze it into packets shaped by a culture of consumers?

To discover some unique and well-supported approaches to answering this very big question, read on…

Don Hanlon Johnson is a professor of Somatics in the doctoral program in Integral and Transpersonal Psychologies at The California Institute of Integral Studies and author of several books, articles, and collections.

Introduction

Notes from the Frontier

The material presented in this book stems from many years spent studying, thinking, and teaching about human potential and the fascinating currents that affect its development. This book and its companion volume, *Anatomy of Embodied Education: Creating Pathways to Brain-Mind Evolution*, contain a distillation of those currents that have aroused my most passionate interest. In the other book, Tim Burns and I have described the processes and functions of the embodied brain vis-a-vis education. In this book I go on to explore how systems-based knowledge about the embodied brain enables humanity both to understand the highest purposes of education and to optimize means of fulfilling those purposes. Finally, I have delved—especially in this book—into the nature of consciousness, mind, intelligence and human development.

This entire book stands as the culmination of my decades-long study of human consciousness. That study comprises a highly rewarding journey of discovery, which now informs and fuels my determination to help inspire a more highly-evolved paradigm of public education that stems from what I have learned. The range and depth of material presented here may be unfamiliar to many readers I would wish to reach, but as another author warned: "The subject is a challenging one and it will certainly require a concentrated effort on the part of the reader." (Edelman, 2004, p. xi). But I would further echo Edelman's caveat as he continued: "I can only promise that the reward for such effort will be a deeper insight into issues that are at the center of human concern" (op. cit.).

My journey of discovery was launched by a singular experience when I was an undergraduate psychology student at the University of New Mexico. The experience was scientifically inexplicable at the time (late 1950s), and is only recently yielding to understanding within the context of neuroscience and neurophilosophy.

The incident followed an intensely emotional evening with Molly, the young woman whom I would wed three years later. I had chauffeured her from Los Alamos, where she lived with her parents, to Albuquerque where she was to board a bus and begin a long cross-country journey. On the way, we had paused to enjoy a romantic dinner at La Fonda Restaurant in Santa Fe. Then we had proceeded to Albuquerque, where we went to a movie and cuddled a lot.

After the movie I escorted her to the Greyhound depot. There, late at night, she boarded a bus for the first leg of a three-week journey that would culminate in a sponsored visit (accompanied by dozens of other students who had won essay competitions in their respective states) to the United Nations headquarters in New York City.

Watching her bus drive away was a heart-wrenching experience for me. Driving afterward through the silent midnight neighborhoods of Albuquerque on my way to my relatives' home (where I had arranged to spend the rest of the night rather than make the two-hour drive back to Los Alamos), I experienced a mixture of loneliness (my sweetheart would be away for weeks!), yearning (I desperately longed to hold her close), and joy (she would return to me!) that I had never felt before.

Such was my emotional state as I let myself into my relatives' house and made my way to their guest room. In that state, exhausted, I fell asleep.

Sometime later that night I awoke in the quiet darkness of an unfamiliar room. Seeing moonlight streaming through a nearby window, I was drawn to get up and look outside. There, cottonwood trees glowed peacefully in the moonlight. Entranced, I felt that peace wash over me. Something unprecedented was happening…

I had no idea where I was, but I had experienced that before for brief moments when waking up in an unfamiliar environment. What was utterly new on this occasion: I had no idea *who* I was! *There was no "I"!* There was only the ecstasy of pure awareness, unfiltered by a sense of remembered identity, unconfined by a limiting ego.

The pure being-ness that I had become stood enraptured for countless moments, joyfully at one with the moonlit trees and sky. Then, little by little, a sense of identity began leaking back in, and as it did the ecstasy began to dwindle, the awareness to narrow, becoming once again conditioned by a sense of myself looking out at trees in the moonlight. It was a lovely scene, yes, but it was not as it had been. *I* was coming back. And with that return came a feeling of regret.

I have spent much of my life since that night seeking to understand what I experienced then, hoping to find a way to re-enter the experience deliberately. For many years the canon of conventional psychology tended to couch episodes like this in terms of pathology (e.g., "depersonalization," "dissociative disorder"). At the same time, wisdom traditions have taught the virtue of subduing the ego. I knew little about either of these poles at the time that I was fortunate enough to have such a state of consciousness arise in me spontaneously. I knew only that I would not stop seeking until I understood the phenomenon.

Several years later, after Molly and I were married, had moved to the San Francisco Bay Area, earned our first tier of degrees, gotten work in our respective professions (education for her, clinical psychology for me), and bought a modest house in the Berkeley Hills, I underwent a second peak experience. A close friend had obtained a 100 mcg dose of LSD and offered to guide me through the journey if I chose to take it.

I did. Without knowing what I know now about the vital importance of "set" and "setting" in psychedelic journeys, I fortuitously or intuitively had ideal versions of both. Tom, my guide, and Molly (who had unknowingly set the emotional stage for my initial peak experience) were lovingly nearby for the duration of my journey. Their presence determined my set. The setting, early in the journey, was the deck of our house on a hillside overlooking the distant Golden Gate Bridge and, beyond it, the horizon of the Pacific Ocean. The air outside was balmy and still, the sky cloudless. The time was just before sunset, and I watched the sun drop slowly toward the horizon, framed perfectly by the Golden Gate Bridge. Just inside the open door leading from the deck into the house, our reel-to-reel tape player poured forth the music of Ravi Shankar.

I stood on the deck watching the sun sink lower as the effect of the LSD intensified, listening to Shankar on tape play an evening raga, with the quiet voices of Molly and Tom audible inside the house. I felt secure in their presence, and knew they were enjoying with me the sunset and the music.

All at once the music and the view fell away, my consciousness sprang outward beyond this ideal setting, and I was aware only of the vast cosmos encompassing the totality of existence.

Time ceased. The music and the sunset seemed somehow woven into the fibers of my being, forming part of the platform from which my awareness had swooped out into the boundless universe, as I became inseparable from all that exists. Bliss, awe, and a sense of ultimate freedom permeated my being.

I slowly came back from that entheogenic peak, as I had from the spontaneous experience in Albuquerque years before. This time, however, it was without the feeling of regret that had accompanied the return from my earlier peak experience.

After that cosmic crescendo, the rest of my LSD journey settled into delightfully novel, intensely enjoyable yet definitely earth-bound perceptual and affective features reported by many other psychedelic travelers. Following a light supper at nightfall, Tom drove us from Berkeley into San Francisco, up Powell Street through Chinatown with its hubbub, clanging cable cars and pulsating lights, out to Ocean Beach to marvel at white-capped waves washing in under the glittering stars, then up to Twin Peaks to look out over the entire lit-up city, then back across the Bay Bridge and home.

All-in-all that journey was orders of magnitude more eventful than the spontaneous, mysterious experience of no-self I had undergone years before. Yet I regard both peak experiences as equally vital to arousing my enduring fascination with fundamental questions about human potential.

Some questions from the initial experience:

- What in the nature of humans gives us the capacity to awaken into such joy?

- What had I needed to transcend in order to experience that blissful absence of identity?

- What conditions enabled me to experience such bliss on that particular occasion?

Some questions from the LSD journey:

- How can human consciousness encompass the magnitude of existence that I experienced at the outset of that journey?

- Is it possible to develop, *without chemical entheogens,* the capability to experience the cosmos to that extent, and to apply that capability to living day-to-day?[1]

- How might humans benefit from applying such capability to mundane existence?

These questions, and others related to them, have fueled my academic, research, professional and personal interests for most of my adult life. My purpose in writing this book is to offer what I currently understand about consciousness and mind/body issues (in the context of complex dynamical systems thinking) to professional educators and even some parents who are motivated and prepared to take in that understanding. I describe the complex processes and functions of the embodied and relational brain, and explore how knowledge about this miraculous brain might enable humanity both to understand the highest purposes of education and to optimize means of fulfilling those purposes.

It is to foster that understanding that the book delves into the nature of consciousness, mind, intelligence and human development.

---

1 Immediately following this initial experience with LSD I intuitively *knew* that I was not to rely on external agents to attain the state I had attained with its help, but that I must learn to enter the state naturally, respecting its power to reveal extraordinary levels of reality. That knowing was accentuated by a few subsequent experiments with entheogens, in none of which did I reach the level I had reached during this initial experience.

I hope to arouse curiosity about these wonders, and to impart enough knowledge of them to contribute to a cultural advance in how we equip young people to grow and flourish.

Aware that I am far from being the first explorer of these realms, I have condensed, compared and synthesized ideas from several other sources I admire. I focus my synthesis through the lens of both classical and contemporary theories about systems. Systems theories—both "general" and "complex dynamical" are extensively described and explicated in Chapters 2 through 4. The value of these theories in helping us to understand the powerful subtleties of consciousness and the embodied brain (that is, the brain-inside-the-skull in concert with its connections to neural networks in the heart and gut) is laid out in Chapters 5 and 6.

I think of systems understanding and neuroscience as a team of two mighty horses pulling in tandem. The payload they haul for me presently is the developmental model unveiled in the final chapter.

That model—originally the brainchild of Tim Burns[2]—is presented in Chapter 7 with an aim toward linking it to the fundamental questions arising out of the peak experiences described above. More importantly from a pragmatic perspective, the model can serve as a blueprint to guide re-visioning the paradigm of education that has prevailed (especially in the USA) for far too long.

To recap, it might be interesting to compare the chapters of this book to parts of a classical music composition. In that case, this introductory section serves as the *overture*, setting out major themes and structures. The initial three chapters then serve to establish a firm foundation for the syntheses to follow. Chapter 1 could be characterized musically as the *introduction*, in which a central challenge is presented and contextualized. Chapters 2 and 3 begin the foundation of a response

---

2   Tim Burns, who has consulted with educators around the world about applying developmental neuroscience to the practice of education, developed the model as a central feature of his work.

to that challenge, involving, in terms of musical composition, *exposition* in Chapter 2 and *development* in Chapter 3. Chapter 4 brings forth the *first recapitulation.*

Chapter 5 furthers the exploration with a *second recapitulation,* and Chapter 6 finalizes the recapitulation cycle. Altogether, these recapitulation chapters, 4 through 6, form a conceptual superstructure arising from the systems-based foundation built in Chapters 2 and 3.

This superstructure consists of summarizing and integrating the published work of several highly regarded systems thinkers/ neurophilosophers who have contributed greatly to my understanding of the relationships among brain, mind and consciousness. When this superstructure—this integrative review—is fitted together with its systems-based foundation. the two components constitute a complex model of emergent intelligence. That is a model that can be applied toward optimal human learning. Our metaphoric composition—the overture you are now reading, the introduction soon to follow, the exposition, the development, the first, second and third recapitulations—might be expected to end there. But the *finale* is yet to come. The complex model of emergent intelligence that has been constructed up to that point is an essential stepping stone to another model—the comprehensive model for developing human potential that is the subject of Chapter 7. And *that* is the *finale.*

The developmental model that constitutes Chapter 7 blends insights from five classical developmental theorists with findings from current neuroscientific research. Taken altogether, these classical insights and contemporary discoveries reveal a multiphase human developmental progression from birth through full maturity, with healthy entry into each phase (beyond the initial one) dependent on fulfilling the potential of the previous phase.

Chapter 7 moves toward answers to the questions posed by my peak experiences in young adulthood. The chapter concludes with a discussion of prevalent sociocultural impediments to full human development all the way from birth through the ultimate phase—termed by Abraham Maslow "self-actualization"—and suggests a vision for the potential role of evolved education in coping with those impediments.

Following the *finale* that is Chapter 7, a *coda* is offered, consisting of a lengthy exploration of how the COVID-19 pandemic can be viewed in terms of systems thinking. Concluding that exploration, I describe an uplifting personal experience that supports a hopeful view of the challenges posed by the pandemic.

Two appendices follow the coda. The first is a discussion of the initial five phases diagrammed in the Chapter 7 Table "Optimal Human Learning and Development." The second is an invited essay by Marika Foltz responding to themes that the book addresses, in terms of her experience both as a student and a teacher.

References

Edelman, G. (2004). *Wider Than the Sky: The Phenomenal Gift of Consciousness.* Yale University Press.

Chapter 1

Applying Systems Thinking to Neuroscience and Education

"Ironically, our various levels of schooling seem to want us to develop our mind, but never train the instrument we'll be developing directly: We pile facts and skills into our children's heads without first aiming an educational spotlight on the mind that is doing the learning."

–Daniel Siegel, MD, *The Mindful Therapist*

A Serious Challenge

The epigraph that begins this chapter contains an implicit and very serious challenge to the prevailing paradigm of education in the Western Hemisphere. The book you are reading stands as one attempt to meet that challenge. I intend nothing less than to promote a radical re-visioning of the narrow, utilitarian paradigm of education so prevalent in the Western world over the past hundred years or so.

The framework I am about to offer is the best one I know of to advance understanding, because it is applicable to systems at every scale—including the socio-political system, the educational system, and the human mind/body system.

In the next chapter I will examine the origins and foundations of general systems theory. In the two chapters following that one I launch an exploration of complex systems, chaos theory, non-linearity, and related topics that have to do with paradigms and how they shift. We must collectively take this on if we are to understand how the institution of public education in the United States became archaic and maladaptive—and understand it deeply enough to be able to rescue it.

If you find yourself flinching at the harsh adjectives applied above to education in the US, let me assure you that I am not the only critic of the current system. Others preceding me include John Taylor Gatto, author of *The Underground History of American Education: A Schoolteacher's Intimate Investigation Into the Problem of Modern Schooling*, and Ron Miller, who wrote *What Are Schools For?*

Those authors have provided clear and comprehensive analyses of systemic flaws in the paradigm that underlies education in the United States. Miller (1997) gives a succinct historical overview of these flaws, describing how, from the onset of the Industrial Revolution when the purpose of public education was bent toward turning out effective and compliant workers, right on through the Cold War from the mid-twentieth century onward—during which the focus shifted to technological supremacy amid harrowing global tensions—the overriding purpose of education in the West has been to serve national interests (determined economically and politically).

As the over-all goal of education was steadily co-opted in the way shown above, the notion of serving the interests of human development and planetary well-being has been systemically marginalized. I aim to put that notion

> The notion of serving the interests of human development and planetary well-being has been systemically marginalized.

firmly in the center of the frame: fostering *true intelligence* in service to life needs to be the primary purpose of education. Other purposes may be in the frame as long as they are compatible with fostering true intelligence, but they must not be allowed to over-ride or obscure its primacy.

What do I mean by "true intelligence in service to life"? And why am I so adamant about advancing it? An extended answer to the first question will emerge as this book unfurls. As a prologue, I will say this: "true intelligence" in humans is supple, wide-ranging, deep-rooted adaptability—both individually and collectively—that gives them the capability to survive and thrive in an increasingly fragile and uncertain environment; "in service to life" refers not just to human life, but to the living environment with which we are deeply connected.

As for my unyielding insistence that fostering true intelligence be the primary mission of education, it is driven in part by my life experience. The peak experiences described in the overture preceding this introductory chapter gave me a bone-deep glimpse into the highest (as far as I know at this moment) potential state of being human. If I can glimpse that state I presume it is possible for virtually anyone else, given the optimal conditions, to reach that state or a higher one. I want that for everyone, and *I believe education can be a powerful creator of optimal conditions for attaining it.*

There is yet another factor in my insistence, a darker one. I share a worry, common among many people I know, that if our culture continues on the path of Business As Usual (ecologically, economically, politically, educationally)—if we do not make a significant widespread shift in our cultural story—we are facing a future that no one could possibly want. This factor is mentioned only in passing a few times throughout the book. I choose not to give it major emphasis, and mention it here only for the sake of completeness; I want the reader to fully understand my motivation to affect positively the educational paradigm prevalent in the developed world.

What is it about that prevalent paradigm of education that needs changing? This is far from being a novel idea, and one objective for bringing forth these complementary volumes (first, *Anatomy of Embodied Education*, and now *Mindleap*) is to amplify the previous critical analyses by John Gatto and Ron Miller, especially in the light of discoveries about brain function and development that have occurred only since the earlier books were written.

In *Anatomy of Embodied Education*, Tim Burns and I have summarized those neuroscientific discoveries that we consider most relevant to our objective. In the current book I discuss from a systems viewpoint the significance of those findings for the mission of optimizing education.

Woven through that discussion, I also examine the discoveries of neuroscience itself through the powerful lens of systems theory. As this examination proceeds, it will shed light on the nature of mind as it relates to neural processes of the embodied brain, of consciousness in general, of what I mean by true intelligence. In the final chapter that examination will culminate in presentation of a unique, comprehensive model of lifelong human development—a model that undergirds my insistence on the need for paradigm change in education.

I will get all of this started with a close examination of Daniel Siegel's declaration in the epigraph at the opening of this chapter.

"Spotlight on the Mind"

Siegel, an eminent neuropsychiatrist, has taken on the challenge of defining and characterizing "mind." He has said that he did this because, after an extensive survey of his fellow workers in psychiatry and psychology, he found not a single one who had attempted this task. Siegel's definition is therefore an excellent place to begin deepening our exploration into consciousness, intelligence, and how these can best be nurtured by educators.

In his introductory lecture at a conference at Garrison Institute several years ago, he began with the question: "What is the connection between the mind and the brain?" This question is obviously of supreme importance to the current exploration, and I report verbatim his explanation of the mind/brain connection:

> The brain is a set of interconnected cells that allow electrochemical energy to flow…When I use the word brain, I mean the embodied nervous system that's distributed throughout the entire body…The mind can be defined as an embodied and relational emergent process that regulates the flow of energy and information…the mind is not just an output of brain function.

He speaks of the mind and brain as two aspects of a single system. A third aspect of that system is relationships, which he defines as the exchange of energy and information. (He speaks of information as "a swirl of energy with symbolic value"). He calls all three—mind, brain, and relationships—"a triangle of human experience," and goes on to elaborate:

> This is talking about three fundamental aspects of one reality. The one reality is energy and information flow. One aspect is the sharing of energy and information flow, called

relationship; another is the mechanism of the flow through the body (i.e., the brain).

> Looking at...the mathematics of systems that are open and capable of going chaotic (we find) something called dynamical laws or complex system laws...What it says is that when there's an open system (that receives something from outside of itself) and it's capable of chaotic behavior... we call it a complex system, which has a couple of principles: one, it's nonlinear (that means small degrees of input could lead to large and unpredictable outputs); ...two, the system has an emergent process...called self-organization. This proposal says that *the mind is the emergent, self organizing process that regulates the movement of energy and information flow.*

Mind, then, is the third aspect of the "one reality" (the flow of energy and information). It is the emergent process that *regulates* that flow, and it co-exists with the other two aspects—relationship and the brain/body.

A brief pause here to draw particular attention to the passage above concerning "dynamical laws or complex system laws." I will be referring back to this quite a bit in the pages to follow.

Siegel continues:

> It's a scientific principle that elements in a system interact—and there's no programmer...[instead of being "programmed"] the interaction of the elements has a property that's called an emergent process, and the emergent process [involves] this triangle that is energy and information flow, and these are three primes of that flow: mind is the emergent process that regulates the flow; relationships are the sharing of the flow; brain is the embodied mechanism of the flow. These are not three different domains of reality; this is one reality.

> ...So the issue is to go deeply into the brain and relationships to understand how this emergent process called mind actually is shaped by this interaction of relationships and the brain.

In his presentation up to this point, Siegel has introduced major facets central to the purposes of the current exploration: open systems, complexity, non-linearity, chaos, and emergent processes. He brilliantly derives his description of "mind" from these concepts, but in this presentation defers thorough explanation of them, due most likely to time constraints. So I will take it on myself to fill in the explanatory background of those concepts, and I will do so beginning with Chapter 2.

I am extremely grateful to Dr. Siegel for the powerfully focused context his presentation provides for the current exploration of mind/ brain/ intelligence within the framework of complex dynamical systems theory (CDST), and will refer to his more recent elaborations on these topics as our exploration culminates.

Now, it is time to fill in the background with a description of systems thinking—why it is necessary, what its basic elements are, and how it has evolved.

References

Burns, T. & Brown, J. (2021). *Anatomy of Embodied Education: Creating Pathways to Brain-Mind Evolution*. Psychosynthesis Press (Originally published by Inspired by Learning, 2020).

Gatto, J. T. (2000). *The Underground History of American Education: A Schoolteacher's Intimate Investigation Into the Problem of Modern Schooling*. Oxford Village Press.

Miller, R. (1997). *What Are Schools For?* Holistic Education Press.

Siegel, D. (2010). *The Mindful Therapist*. W. W. Norton & Company.

Siegel, D. (2011). The Neurological Basis of Behavior, the Mind, the Brain and Human Relationships. Climate, Mind and Behavior Symposium, Garrison Institute. https://www.youtube.com/watch?v=B7kBgaZLHaA

Chapter 2

A Bit of History About Systems Thinking

What Led Me into the Study of Systems?

My exploration of the mysteries of consciousness began in the early 1970s during the interim between walking away from the first doctoral program I had enrolled in—at the Berkeley campus of the University of California—and four years later entering my second program at another UC campus, UC Santa Barbara. During those interim years my wife Molly and I dedicated ourselves to parenting our two young sons, scraping together meager earnings from leading encounter groups, and reading the burgeoning literature on humanistic psychology, altered states of consciousness, and human potential in general.

That trajectory continued after I enrolled in the doctoral program in counseling psychology at UCSB. My mentor there was Stewart Shapiro, an ebullient, creative teacher of what he termed "the psychology of positive experience." He had been the chairperson of the Counseling Psychology Department at UCSB when I applied and was accepted the year before. By the time I arrived, though, he had been replaced as chair by a dedicated behavioral psychologist—and my continued participation in that program was doomed. Behaviorism was what I was determined to leave behind. I withdrew at the end of the Spring Quarter, less than a year after landing there.

Nevertheless, during those three quarters at UCSB I learned much from Stewart Shapiro. He was one of the most accessible teachers I ever encountered, subtly exuding an atmosphere of excitement about the knowledge and experiences he had gathered, along with a generous eagerness to share what he had with his students. Just being in his presence constituted education in its etymological sense (*educere*: Latin, to draw forth).

Yet the Counseling Psychology program itself did not satisfy me, and I took another leave of absence that never ended. Even while I was still enrolled there, Molly and I began our extra-curricular involvement in an emerging field of knowledge and methods known as psychosynthesis. This is a transpersonal psychology originated by Italian psychiatrist Roberto Assagioli. Begun in Europe, it had gained traction in Canada and the US—especially in coastal California. The subject matter was precisely what I had been seeking—and not finding—in my doctoral studies thus far.

We both joined the first cohort of Bay Area trainees in this exciting field, and would later travel to Italy to be tutored directly by its founder. Molly eventually wrote and published two books about psychosynthesis, and is still acknowledged as a pioneer in that community.

I, on the other hand, attended psychosynthesis trainings for a few years, but my interests characteristically burgeoned within the ever-expanding field of transpersonal psychology. Within a few years psychosynthesis was to be joined in my assortment of resources by methods and concepts which, at first glance, bear little resemblance to each other. Yet for me their resemblance is obvious: every field of knowledge I pursued swirled around the tantalizing mysteries of human consciousness. "What is our purpose and potential?" I wanted to know, "and how can we best fulfill it?"

My arc of discovery had been boosted to a higher orbit immediately after I withdrew from my doctoral studies at UCSB (and the irony of that does not escape me). During the summer after leaving that program, in addition to linking with the nascent psychosynthesis community in the Bay Area, I attended a 2-month residential program, sponsored by Esalen Institute and held on the campus of Stanford University. The subject matter was the study of human consciousness. One of the organizers of that program, James Fadiman, was on the faculty at Stanford. Another organizer was Robert Ornstein, a neuroscientist at UC San Francisco. These two pioneering researchers/scholars had assembled a stunning array of teachers (including Charles Tart, Arthur

Deikman, Claudio Naranjo, Lawrence LeShan and Robert Frager) to join them in presenting cutting-edge theories and discoveries beyond any I had yet encountered.

That intensive 8-week program was the most exhilarating learning experience of my life up until then, and it definitely set the mold for my subsequent life of learning. Fatefully (and now we come to the crucial point of this lengthy autobiographical interlude) it also afforded me my initial experience of, and basic instruction in, biofeedback training. That was my introduction to neuroscience, and it led me inevitably to the study of systems.

Within a few years of that life-changing summer I would at last enroll in the doctoral program of my dreams at the Humanistic Psychology Institute (now known as Saybrook University), where I could study humanistic and transpersonal psychology—the psychology of *consciousness*—to my heart's content, under the tutelage of some of the most eminent figures in those fields. Conversations, courses and correspondence with James Bugenthal, Stanley Krippner, Marcia Salner, Don Polkinghhorne, Dennis Jaffe and many of their peers challenged and delighted me.

The culmination of my study there, my dissertation research, was to test a theory I had gleaned about the relationship of brainwave patterns to carefully measured states of consciousness. My theory necessitated familiarizing myself with general systems theory as well as neuroscience. Those two disciplines are intimately related, and have occupied my attention ever since I first encountered them. What follows reflects my current understanding of both.

General Systems Theory arose in the mid-20th century as some pioneers in science began realizing that current methods and models in science—particularly biology—were painfully inadequate in explaining the nature of the living world. These pioneers, particularly Ludwig von Bertalanffy (1968), set out to create new ways of looking at and thinking about the world around us and within us. They recognized the folly of dividing the phenomenal world into smaller and smaller parts for analysis, an approach now called reductionism. The assumption underlying the reductionist approach is that by breaking something down to its constituent parts, you can understand that something.

In certain ways the reductionist approach resulted in spectacular technical advances when applied to mechanisms of various kinds. But, as von Bertalanffy and others were starting to realize, reductionism did not lead to deep understanding. What was missing was the study of how all the parts of something, particularly an organism at any level, function *together*, how those parts interact with each other as a whole thing, a system. This realization led to a focus on systems of many kinds, and ultimately to formulating the principles that underly the functioning of systems in general. Knowledge of those principles turns out to be a powerful tool for understanding and working with the world of phenomena, the world we experience every day.

What is a System?

We live in a world of systems. Indeed, we use the term "system" quite a bit, referring to specific systems (such as the heating system in a building, the cooling system in a car, a body's digestive system, the next storm system heading our way) without thinking much about what systems comprise in a general way. Ordinarily we get by without feeling the need to understand the general nature of systems, but in the current project we must look deeper.

We know we live in a solar system, homes have heating systems, cars have electrical systems, bodies have digestive systems. What constitutes a system?

Think of a set of elements that all connect together, each one affecting all the others. They all function together as a whole, and that whole accomplishes something—keeps a home at a comfortable temperature, generates and distributes electrical current to the parts of a car that need it, extracts energy and building material from food and excretes waste—those sorts of accomplishments.

Holarchy

One of the most fascinating things about systems is their holarchic nature. What that means is that each element in a given system is itself a system (like the alternator in a car's electrical system or the stomach in a body's digestive system), and each system is itself an element in a more comprehensive system (the entire car, for instance, or the entire body). In writings about living systems, you will find the term "nested hierarchy" referring to this perspective (i.e., systems nested within higher systems almost never-endingly). In his book *The Ghost in the Machine* (1968), Arthur Koestler coined the word "holon" to mean both whole and part. "Nested hierarchy" then becomes "holarchy," or even "holonarchy" (p. 48). Macy and Brown[3] also rely on this concept.

Living and Non-Living

The "living systems" term implies, correctly, that some systems are alive and some are not. What differentiates living systems (synonymous with natural systems) from non-living or artificial systems? I introduced, examples of each above—bodies and cars—and come now to the distinction as systems theory formulates it.

Remember that systems, in the more-or-less technical sense used here, always accomplish something. Now, as far as anyone knows, nothing in the universe gets done without some form of energy being involved. That means systems need energy to do what they do. Where does this energy come from? Here's where we get to the difference.

A living system is open to its environment, proactively taking in energy from other systems around it (the biosphere) in a chain tracing all the way back to the sun. It processes those inputs of energy (commonly known as food or fuel), deriving from them a form of energy it can use to do its thing, then releases what it does not use back to the biosphere

---

3   "Every system is a 'holon'—that is, it is both a whole in its own right, comprised of subsystems, and simultaneously an integral part of a larger system. Thus holons form 'nested hierarchies,' systems within systems, circuits within circuits, fields within fields." (Joanna Macy and Molly Brown, *Coming Back to Life*, New Society Publishers, 2014, p. 40.)

to be used as food for some other system. All along the way—input, throughput, and output—the living system functions autonomously, meaning it sustains itself. All its subsidiary and overarching systems, all holons, do the same thing. In no case, however, does "sustaining itself" mean "in isolation." It is just the opposite: living systems are open and inter-connected. Inter-connectedness is essential to living systems.

In sharp contrast, a non-living system is closed off from its environment, and must have energy artificially provided to it before it can function. It cannot proactively obtain energy for itself. It produces waste that seldom serves as fuel for other non-living systems, but instead increases disorder in its environment. In brief, non-living systems do not sustain themselves.

Let's look more closely at how a living system sustains itself. That process turns out to have two distinct and vital facets: self-regulation and self-organization.

Self-Regulation

A living system such as a human body regulates its own internal environment within a shifting external environment. For example, one's body keeps its temperature within a livable range by (among other processes) moving warm blood to the surface when excess heat needs to be radiated to the environment, or redirecting blood to the core when it's cold outside and body heat needs to be conserved. How does your body know when to do which? By relying on sensory networks, which give *feedback*—information about how environmental conditions are affecting the body and how well the body is adjusting to them.

Without this information, the body would have no basis on which to coordinate its myriad subsystems so that it and all its constituent holons survive.

Systems thinkers make extensive use of loop diagrams (also known as "causal loop" diagrams) in applying information-response dynamics to real world situations. An example of such a diagram is shown in Figure 2-1, which illustrates the physiological loop just described.

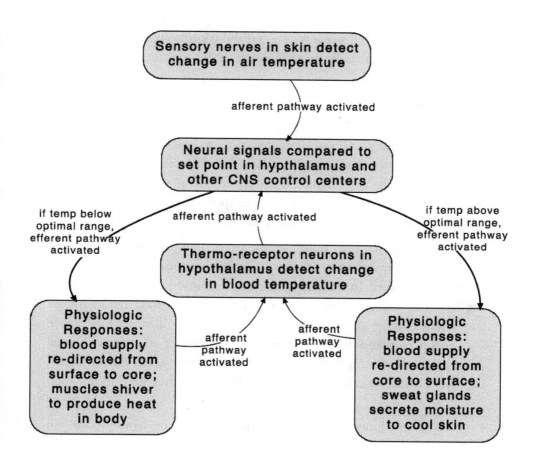

Figure 2-1. Among the ways the bodies of warm-blooded animals stay within an optimal internal temperature range is autonomic or self-regulating, using the kinds of feedback loops illustrated here. In addition to these internal processes, behavioral changes may also be used to adapt to changes in external temperature, and these could also be diagrammed as causal loops and incorporated with the ones shown above. (Diagram by Jim Brown)

The kind of feedback that enables living systems to self-regulate is called deviation-reducing, or simply negative, feedback.[4] It enables the system to reduce the deviation between the conditions favoring survival and the conditions actually prevailing within the environment.

Self-Organization

In addition to being self-regulating, living systems are also said to be self-organizing. These characteristics are both distinct from each other and complementary, each to the other. We will look now at what that means, for the two characteristics in tandem are at the core of living systems. As we shall see in the next chapter, they are also crucial to understanding complex dynamical systems.

Returning to the example of body temperature relative to environmental conditions, imagine a primitive human who has lived comfortably naked in a temperate climate. Now imagine the climate suddenly becoming so much colder that the person's body can no longer stay warm enough through the internal processes illustrated in Figure 2-1. This person and his or her companions, who are all facing the same crisis, might discover quickly that huddling together keeps them warmer. But they cannot remain huddled and still do what they need to do, such

---

4   When used to distinguish types of feedback in systems, the terms negative and positive do not carry value connotation as they do in more colloquial contexts. In a "negative" feedback loop, a quantitative increase in the output of one holon induces a quantitative *decrease* in the output of other holons to which it relates. These decreased holons then feed back signals to the originating increased holon that result in damping down its output. (A shorthand version of this process would be "the more *x*, the less *y*, leading back to less *x*.") Overall, this has the effect of maintaining equilibrium within the system.

A positive feedback loop, on the other hand, is one in which a quantitative increase in the originating holon induces a quantitative *increase* in holons that it affects around the loop, which then signal the originating holon to keep on increasing its output. (A comparable shorthand version of this process would be "the more *x* the more *y*, leading back to more *x*.)* The overall effect of this is to push the system increasingly to the limits of its viability.

Negative and positive feedback thus translate to the more precise terms deviation-reducing and deviation-amplifying feedback, where "deviation" means departure from a systemic norm or code.

as obtain food. Let us suppose they have observed that other mammals are covered with fur and live in dens. These humans emulate this by finding ways to cover and shelter their own bodies, along with adopting the use of fire for warming themselves and cooking their food. By doing so they manage to survive. They have adapted by learning effective ways to change their behavior, leaving behind the old status quo and reorganizing around a new one. Deviation-reducing feedback will resume, but at a higher level (behavioral/cultural) that incorporates the previous level (physiological).

Just what does it mean for deviation-reducing feedback to resume at a higher level, following an adaptation brought on by a critical change in the environment? Grasping the significance of this pattern is crucial, because we are describing nothing less than a core feature of evolution and learning.

> Grasping the significance of this pattern is crucial, because we are describing nothing less than a core feature of evolution and learning.

The example just given, humans changing cultural behavior as an adaptive response to environmental change, is only one of myriads of examples that I could have chosen from the history of human evolution. Weaving through every one of those examples would be the systems principles addressed in this chapter: that using deviation-reducing feedback, systems self-regulate within existing environmental parameters, and that when those parameters change (as they inevitably do), systems survive only by shifting to a higher level of complexity to accommodate the change in parameters. A shift like this is known in general systems terminology as self-organization.

Following successful adaptation to environmental shifts (for that is what self-organization is all about), systems settle back into deviation-reducing feedback dynamics suited to their changed environment. But what if the adaptation doesn't "take" for some reason? What if, instead of evolving to a level suitable for an altered environment and then reinstating a regime of deviation-reducing feedback that works in that environment, the system gets caught in "run-away positive feedback?"

In such an event, wherein a system faces unchecked positive feedback, there is a danger that it will disintegrate chaotically instead of breaking through to an advanced level of adaptation. In the following chapters I will suggest ways in which our system of education—and the individual learners who can be considered holons of that system—need to be guided to self-organize at a level of complexity requisite to meet the complex challenges of our changing world, while at the same time avoiding the risk of disintegration. To aid us in that exploration we will, in Chapter 3, call upon a systems approach that itself has evolved from the general system theory just described.

> Our system of education—and the individual learners who can be considered holons of that system—need to be guided to self-organize at a level of complexity requisite to meet the complex challenges of our changing world.

References

Koestler, A. (1968). *The Ghost in the Machine.* The MacMillan Company.

Macy, J. and Brown, M. (2014) *Coming Back to Life.* New Society Publishers.

von Bertalanffy, L. (1968). *General System Theory.* Braziller.

Chapter 3

The Next Level of Systems Thinking— Complex Dynamical Systems

In the decades since Ludwig von Bertalanffy and others began articulating and advocating basic principles of systems thinking, the field has inevitably grown wider and richer with the exploration, for instance, of how complexity characterizes and affects systems. Neuroscientists persistently use the word "complex" as a descriptor for the human brain. In the current chapter I intend to circle in more closely on how complexity ties in to evolving systems thinking.

I graduated from Saybrook without any inkling of how systems theory was about to evolve. Soon after graduation Molly and I moved from New Mexico, returning once again to northern California where I figured my long-sought doctoral degree would be most useful to me. That move enabled me to meet Jon Klimo, an interdisciplinary live wire who became very interested in the research I had done for my doctorate linking non-ordinary states of consciousness to specific brainwave patterns. Jon was writing a book on a particular form of non-ordinary consciousness known as channeling, and cited my research in his book.

In 1987, a year after Jon's book was published, he invited me to participate in a day-long seminar he had organized, titled "A Discussion of Interdisciplinary Connections toward the Integration of the Sciences" (jovially subtitled *A Day for Playful Minds*). Among the fourteen other invited participants was Ralph Abraham, a professor of mathematics at UC Santa Cruz who was working on what was then known as chaos theory—which has since evolved into "complex dynamical systems theory (CDST)," the primary subject of this chapter.

Hearing about it that day from one of its pioneering theorists was my first exposure to this way of understanding systems. I corresponded briefly with him, and he sent me several issues of his periodical *The Dynamics Newsletter*. Soon afterward I discovered James Gleick's *Chaos* (1987), which furthered my absorption in this revolutionary field of study. That book swept me through the history and substance of dynamical systems theory. In it, Ralph Abraham's characterization of the burgeoning field is quoted: "It's the paradigm shift of paradigm shifts" (p. 52).

My absorption became dormant, however, as I had no direct way of applying CDST to my work in biofeedback training and counseling. Pursuing that work drew me deeply into the study of general systems, especially pertaining to feedback loops within systems. Understanding feedback and its intricacies did have immediate practical application to my work in biofeedback. Within the general field of biofeedback, my primary interest gravitated toward neurofeedback—a natural progression given my fascination with consciousness in general.

Thus, understanding the workings of the human brain became, for me, inextricably linked to understanding systems in general. But it was not until my longtime friend and associate Tim Burns invited me to help him complete a book compiling a set of teachings about the brain vis-a-vis education that I was driven to revive my study of complex dynamic systems. In revisiting that field I came at last upon my version of the "eureka" experience, finding the pattern that brought my long sought grasp of consciousness, the brain, mind-body unity, into focus.

It has taken me years to assimilate complex dynamical systems concepts well enough to attempt writing about them. And I still would not dare to make the attempt if there were any other adequate way to encompass the subject matter of this book: furthering the highest purposes of education, and of human development in general.

I will begin by looping back to the concepts introduced earlier by Daniel Siegel: open systems, chaos, complexity, non-linearity, and emergent processes. I know from personal experience that these are difficult concepts to grasp. Yet they cannot be avoided if we are to comprehend fully:

- the relationship of mind and brain,
- the relationship of mind/brain to learning,
- what intelligence means in the context of that relationship,
- and what needs to happen in schools to optimize learning and intelligence.

These matters are of crucial importance to all of us humans as we seek optimal ways to survive and thrive in perilous times. If you agree with that assessment, I hope you will also agree that comprehending these concepts is well worth the effort required to do so.

Complexity

We already have a start in understanding the meaning of complexity, as far as systems are concerned, from an earlier definition of a system: "a set of elements that all connect together, each one affecting all the others." Basically, complexity in a system is a function of how many identifiable elements the system comprises, and how those elements connect with each other. As you might have already read in other sources, human brains contain some hundred billion neurons, with each neuron synaptically connecting to many thousands of others. That is why it can be claimed that the brain is a system of astronomical complexity (and that is without even considering the enormous complexity within each of those neurons, or the complexity of brain-body interconnections).

There is much, much more to be said about complexity. In the scientific literature reviewed for these two books, complexity crops up again and again. Often it is inseparable from the other terms noted above (i.e., open systems, emergence, and especially non-linearity.) So let's begin to explore the wide range of issues around complex systems, in the spirit of interweaving them into a wide and comprehensive pattern of understanding.

Prize-winning neuroscientist György Buzsáki, whose monumental book *Rhythms of the Brain* was a major source of material for this one, gets us off to an exemplary start:

Complex systems are open, and information can be constantly exchanged across boundaries...Oftentimes, not only does complexity characterize the system as a whole, but also its constituents (e. g., neurons) are complex adaptive systems themselves, forming hierarchies at multiple levels. All these features are present in the brain's dynamics because the brain is also a complex system.

...These complex systems live by the rules of nonlinear dynamics, better known as chaos theory. (p. 11)

Buzsáki explains further:

Complexity can be formally defined as nonlinearity... Because the constituents are interdependent at many levels, the evolution of complex systems is not predictable by the sum of local interactions. (pp. 13-14)

He later reiterates:

Complexity arises from the interaction of many parts, giving rise to difficulties in linear or reductionist analysis due to the nonlinearities generated by the interactions. Such nonlinear effects emerge from both positive (amplifying) and negative (damping) feedbacks, the key ingredients of complex systems. (p. 54)

It is important to notice in the paragraph quoted above that Buzsáki draws a solid connection between the general systems model, discussed in the preceding chapter, and the more recently derived model of complex dynamical systems. By linking positive and negative feedbacks to emergence, non-linearity, and complexity, he clarifies succinctly how the earlier model provides the foundation for the later one.

Non-Linearity

Buzsáki's phrase "linear or reductionist analysis" needs to be examined more closely, because it relates to the pejorative adjectives used near the beginning of Chapter 1 to describe the prevailing paradigm of education in the United States and beyond. I called the paradigm archaic and maladaptive. Before proceeding further in this tutorial on complex systems and nonlinearity, I need to explain why I have such a critical attitude toward the prevailing paradigm.

I also need to make it very clear that my criticism is about the paradigm, not about the educators who are caught up in it. I am certain that there are many of you who feel as I do, that the prevailing paradigm is archaic and maladaptive, and who recognize that it needs to evolve. Yet despite that wide recognition, despite the scathing criticisms that have been written about it, despite even the many positive alternatives to it that have been proposed[5], the paradigm has not yielded its predominance.

It is difficult to know precisely why it is so entrenched, but the entrenchment is undeniable. All my life I have witnessed stubborn resistance to radical change at levels of systems ranging from personal to cultural. Toward the cultural end of that spectrum, within the institution of education, many factors may be found contributing to the resistance against adapting its paradigms and methods in the ways proposed in this book.

It is my fervent hope that these factors may ultimately be understood and corrected. In this next section I address what I feel to be one of the basic factors in that institutional resistance: unquestioning adherence to a linear and reductionist world view.

Linearity and Reductionism: A Critique

Linearity is generally assumed in our culture to be a predominant feature governing the observable universe. For example, a force applied to an object will cause the object to move (think of kicking a soccer ball); the velocity of the object is generally assumed to be in direct proportion to the amount of force applied. The greater the force the greater the velocity. That is a linear function. Newtonian Physics is built around such functions, and Newtonian Physics is still widely considered the governing paradigm of science at the macrocosmic level.

---

5   See for instance Gang, Meyerhof & Maver, *Conscious Education: The Bridge to Freedom*, Vermont:Dagaz Press, 1992.

That paradigm has worked quite well in the so-called hard sciences, physics and chemistry, although even there it is increasingly challenged. The so-called soft sciences, such as psychology, sociology, biological and health sciences, and education, have embraced reductionism, along with its underlying assumptions of linearity, apparently without closely examining whether these assumptions are truly applicable to them.

Marc H.V. Van Regenmorte (2004) is an apparent exception, exemplifying a growing tendency among scientists to question the reductionist paradigm when he wrote:

> The reductionist method of dissecting biological systems into their constituent parts has been effective in explaining the chemical basis of numerous living processes. However, many biologists now realize that this approach has reached its limit. Biological systems are extremely complex and have emergent properties that cannot be explained, or even predicted, by studying their individual parts. The reductionist approach—although successful in the early days of molecular biology—underestimates this complexity and therefore has an increasingly detrimental influence on many areas of biomedical research, including drug discovery and vaccine development.

Joe Lyons Kincheloe (2005) was a major trail-blazer in the field of critical pedagogy until his death in 2008. He had this to say about reductionist trends in education:

> Contemporary efforts at educational reform with their specified facts to be learned for standardized tests handcuff teachers as they attempt to teach complex concepts and to connect them with the needs and lived experiences of students. In this crazy context such educators are victimized by a simplistic and frightened response to social change, youth-in-crisis, or a decline in standardized test scores. Putting their faith in reductionistic measurements of student memorization of disparate fragments of data, advocates of No Child Left Behind (NCLB) and other top-down reforms have no basis for evaluating more sophisticated dimensions of

learning, thinking and teaching. Indeed, they can't measure even the traditional skills of good scholars, not to mention the innovative and evolving operations of intellects coming from diverse cultures and differing paradigmatic (meaning frameworks for making sense of the world) perspectives. As I have argued elsewhere...even the world-changing scholarship of an Albert Einstein in physics—portions of which, such as the Special Theory of Relativity, are almost a century old—cannot be taught, learned, or evaluated in the educational swamp of imposed, right-wing reform. (p. 85)

More than a decade has gone by since Van Regenmorte and Kincheloe wrote their indictments of reductionism in education, and still the situation is a dire one. More recently, Henry Giroux (2016) was driven to publish an even more biting critique:

America's obsession with metrics and quantitative data is a symptom of its pedagogy of oppression. Numerical values now drive teaching, reduce culture in the broadest sense to the culture of business, and teach children that schools exist largely to produce conformity and kill the imagination.

Note that, although he begins this statement speaking specifically about the obsessive reliance on standardized testing and the quantitative data it generates to determine school policies, Giroux quickly sweeps into the heart of our viewpoint in this book: that public education in the US (and elsewhere in authoritarian-leaning societies) has generally devolved to the point of existing "largely to produce conformity and kill the imagination."

In brief, policies in school systems throughout the US and other industrial countries evidently favor the factory model that dates to the time of Henry Ford: the purpose of schooling is to produce workers inured to rigid schedules, lock-step procedures, relentless competition and stress- saturated work environments. This model of education is at best regressive, and at worst inhumane.

Please remember—if you are a teacher reading the above indictments—that they are not directed at you as an individual. That is the opposite of my intention. The point is that, under the prevailing paradigm of education in North America, teachers and students alike are victims of the "pedagogy of oppression."

Back to What Works

The above critique documents a growing realization that education in developed nations such as the US has hitched its wagon to a fading star: a linear, reductionist paradigm that is losing favor even in mainstream science. Let us now proceed in our project of understanding the value of shifting to a paradigm that recognizes our world as fundamentally complex and nonlinear. The part of the world that this book focuses on, the embodied brain/mind/consciousness of human beings as impacted by our system of education, is particularly and potently imbued with complexity and nonlinearity.

The case being made here is that neuroscience discoveries about brainbody structure, function, and development must be taken into account in designing and delivering educational programs. To back up that case, it becomes necessary to delve more deeply into issues introduced earlier, issues that command an increasing level of attention from neuroscientists and neurophilosophers.

> The case being made here is that neuroscience discoveries about brain/body structure, function, and development must be taken into account in designing and delivering educational programs.

Let's go over those issues again, succinctly, in a list format:

1. Understanding the structure, functions, and development of the embodied human brain;

2. Understanding the relationship of consciousness to the brain's structure, functions, and development;

3. Defining the relationship of consciousness to mind;

4. Exploring the roles of attention and intention in how that relationship plays out;

5. Suggesting the purpose and nature of intelligence;

6. Evoking and applying a model subtle enough and powerful enough to encompass all of the above, thus enabling humans to better understand and cope with our world.

The first of these issues has been thoroughly addressed in *Anatomy of Embodied Education*. Issues two through five have been repeatedly alluded to, and will presently be explored in depth. Number six, creating a subtle and powerful model, has been the primary subject of the current chapter to this point.

Introducing...The Model!

A model, in the sense used in this chapter, is a mental construct that represents the phenomenal world within and around us. Its purpose is to enable clearer, deeper understanding of our world so that we may interact more effectively with it. In my view an outstanding model for this purpose is complex dynamical systems theory, CDST.[6]

In the following pages I will summarize the major features of CDST that enable conclusions about the six issues listed above.

You have already been introduced to some of these features in the quotes from György Buzsáki's book, *Rhythms of the Brain,* in which he indicated that complex systems are nonlinear, interdependent at many levels, and that they function far from equilibrium.

That is just the beginning. Complex systems are also characterized as dynamic—meaning that they change over time. To understand those systems better it helps enormously to identify the patterns that underlie how they change. The CDST model used by many scientists to understand complex systems employs ingenious mathematically-based methods for plotting and visually illustrating those patterns. We need not get into the mathematics involved, but we will need to focus more closely on a few aspects of the visual illustrations in order to grasp the

---

6   Adding the suffix -al to dynamic indicates that the systems being studied are capable of being modeled mathematically. In this book the two terms are considered interchangeable.

fascinating potential of this model to enable deeper insights into our complex and often bewildering world.

I want to introduce some aspects of graphing and illustrating typical patterns that systems follow as they change—aspects that are most crucial to our exploration. Chief among these is the notion of attractors, which generally refers to a tendency for the system to change as though being pulled or guided by a force of some kind.

In the current context the general term "pattern" must be narrowed down to something known as "trajectory" in CDST. This refers to a visual representation of sequential measurements of variables, with each variable representing a particular holon (a concept introduced in Chapter 2) as the state of that holon changes over time. In other words, a trajectory in this context is a graph of change of some aspect of a system over a given period of time.

Concrete examples will be given later of the very abstract terms being introduced in this segment. For now I will just say that, in general, observation and measurement of several such trajectories enables the observer to infer which type of attractor is affecting the system being observed. That inference can in turn enable the observer to understand *how the system is likely to evolve.*

The three images pictured below are idealized illustrations of trajectories under the influence of three general types of attractors: fixed point, limit cycle, and chaotic. In each image you will see lines with directional arrows, which represent flowing trajectories of holons within complex systems changing through time.

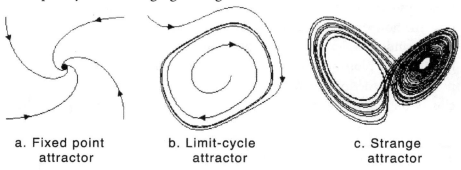

a. Fixed point   b. Limit-cycle   c. Strange
   attractor   attractor   attractor

Figure 3-1. Three General Types of Attractors

The trajectories depicted in all three images are composed of sequential measurements of variables, with each variable representing a particular holon as the state of that holon changes over time. For instance, a system governed by a *fixed point attractor* (*a,* above) will ultimately reach equilibrium and cease changing. A simple example of such a system is a wind-up clock, which ticks until the energy stored in the spring that drives the clock's hands ceases to exceed the forces of gravity and friction, at which point the clock stops, having reached a static sort of equilibrium. Such a mechanism meets the criteria for a system, but in the terms discussed in Chapter 2 cannot be regarded as a *living* system. Why not? Because it is closed off from its environment, and must have energy artificially provided to it before it can function.

The *limit cycle attractor* (labeled *b* in the illustration) will cause the system to oscillate within a fixed range of values. Now we *are* talking about living systems. A classic example of such a system in physiology is the pulsing of the circulatory system. Another is the rhythmic firing and resting of synchronous brain cells in a neural net that gives rise to the signals measured and displayed on an EEG (electroencephalogram) instrument (whether it be in a hospital, a research laboratory, or a neurofeedback clinic). Each of those holons, portrayed in the dynamic way shown in *b,* would be seen to cycle repeatedly within certain parameters—thus the term limit cycle attractor.

The third type of attractor, exemplified by the *Lorenz attractor* shown on the right in Figure 3-1, is the most mysterious and arguably most ubiquitous and compelling of the three. It is the kind of attractor that brings "bifurcation" into play. Looking closely at the butterfly-shaped Lorenz attractor you can see that the trajectory switches back and forth between what appear to be a couple of limit cycle attractors (similar to the one shown in the center), apparently never coming to rest in either one, but rather linking them together in a strange way. In fact, the word "strange" has come to be associated with attractors like this.

It is the "switching back and forth" that illustrates what bifurcation means: the forking of a system's trajectory when one attractor ceases to hold sway over the system, either by ceasing to exist or by losing out to a more compelling attractor.

Bifurcation can be equated with radical change, transformation—and all these terms point to the heart of what is meant by evolution.

To bring these abstract terms into a real-world example of a complex dynamic system, think of a musical performance that you have experienced. A jazz ensemble performance would work very well for this thought experiment, largely because of the improvisational nature of jazz. A jazz number often begins with a recognizable melody, after which various players take turns improvising an interpretation of the melody, or theme. If each player's series of musical notes were to be scored, that score could be thought of as a trajectory of notes, and the theme melody around which the players build their improvisations would represent an attractor for all those trajectories. The performance of the theme melody and its variations is thus a complex dynamic system.

A jazz number often begins with a recognizable melody, after which various players take turns improvising an interpretation of the melody, or theme. If each player's series of musical notes were to be scored, that score could be thought of as a trajectory of notes, and the theme melody around which the players build their improvisations would represent an attractor for all those trajectories.

Now, to illustrate what is meant by the novel term "bifurcation," let's imagine that the chosen theme melody is similar in some ways to another melody known to all the players, and that one of the players begins to riff, either deliberately or not, around this alternative melody (I have actually heard this done, and was entertained by the surprise it brought on[7]). The other players would no doubt notice this, and might

---

7 A more deliberate instance of this jazz phenomenon can be found in a 1959 recording: Dave Brubeck Quartet - "Blue Rondo à la Turk," live - YouTube. The number opens with a lengthy, extraordinarily complex segment in 9/8 time with an exotic melody line, shifting suddenly into a cool, swinging, more traditional-sounding segment in 4/4 time (toward the end of which tantalizing hints of the opening segment are inserted), concluding with a full-on shift back into a reprise of the opening segment. Brubeck was experimenting with and demonstrating bifurcation decades before it became a topic in CDST!

choose to bend their improvisations toward this alternative theme. Such an occurrence would constitute a bifurcation in the performance—a shift in trajectories toward a competing attractor (in this case, the alternative melody).

Here's another example that is closer to the subject matter of this book; it is a subtle bifurcation that every reader of this book has undergone, whether they noticed it or not.

Among the contents of *Anatomy of Embodied Education* is a description of the autonomic nervous system (ANS), which has two branches: the sympathetic branch and the parasympathetic branch. During most of a person's waking hours the sympathetic branch tends to predominate, especially if that person's life is busy and full of challenges, large or small.

All of us must sleep sometime, but when we lie down to sleep the onset of sleep can be delayed if the sympathetic branch of the ANS is still firing away. At some point, if sleep is to come, a shift must take place from sympathetic to parasympathetic dominance. That shift can be noticed by people who are mindful of their physiological and mental processes. Among other phenomena, the shift can be signaled by an abrupt slowing and deepening of one's breathing, a rapid release of skeletal muscle tension, and, more subtly and occasionally, awareness of hypnagogic images that can arise with one's eyes closed.

Anyone fortunate enough to have noticed such phenomena has received the gift of being aware of a subtle bifurcation, a phase shift in which one attractor, the functions of the sympathetic branch of the ANS, suddenly gives way to the functions of the parasympathetic branch—the alternative attractor—enabling sleep to come and cerebral repairs and housekeeping (which are among the recognized purposes of the sleep state) to begin.

As pointed out earlier, complex, dynamic, living systems *change* as time goes on (thus the term "dynamic"). Often—more often then you might think—the change is dramatic. The above examples are intended to give a sense of how one kind of change, bifurcation, occurs, and what the system becomes as a result of the bifurcation.

Yet Another Step

To move our understanding further, one more step must be taken. Recall the discussion in Chapter 2 about self-organization in living systems. Frederick D. Abraham (2014) guides us back to that discussion at a new level:

> Self-organization refers to the fact that one or more control parameters of a system are a function of the state of the system and thus the behavior of the system will change accordingly. The changes may be subtle or gradual, or if a bifurcation point is passed, they may be dramatic…Self-organization is where a system can create changes, including bifurcations, by influencing its own control parameters… (p. 15)

There is much to unpack in the above, critically important, passage. The key term is the admittedly arcane phrase "control parameter," which hardly anyone but an engineer or a scientist would ever use. However, it is necessary for us to grasp it as it relates to dynamical systems modeling, in order to bring our understanding to fruition. That fruition includes clarity about how emergent properties spring up in complex systems—novel properties that could not be predicted just by knowing the systems' components. And knowing about emergent properties is vitally important to understanding how learning and evolution proceed.

Knowing about emergent properties is vitally important to understanding how learning and evolution proceed.

Many different definitions of "parameter" can be found, definitions that shift according to the type of system to which the word is applied (e.g., biological, physical, sociological, even musical). Here I attempt to gather and summarize the commonalities among several of those definitions: *a parameter is a measurable element, quality, or combination of elements, qualities, etc., that help drive, define or otherwise characterize a system or any holons of a system.*

If that definition brings on mental indigestion, welcome to the club. Some concrete examples might help to soothe your indigestion.

For one example, think of a weather system. Is it going to rain or snow today? Forecasters attempt to answer this very important question by measuring atmospheric pressure, humidity, and related variables of your particular region. If the atmospheric pressure is low enough to allow moist air to flow into the region, the moisture could precipitate and fall to the ground in the form of water (rain), ice (hail or sleet), or something in between (snow). What determines the form of precipitation? The temperature of the air does.

In this system, air temperature is considered a parameter. It is a variable that can be measured, it rises and falls independently of the humidity level, and its numeric value determines whether precipitation falls as rain or snow, so it is a *control* parameter.

As you mull over the example above, it might occur to you to wonder: if changes in air temperature (named here a "control parameter") determine the form of precipitation, then what determines—controls—those changes in air temperature? That is a legitimate question, and it conveniently leads us to a subtle feature of parameters that affect complex systems: those parameters can themselves be affected by the systems that they influence—as Frederick Abraham (2014) states at the end of the passage quoted on the previous page. (If this reminds you of the causal loop diagrams discussed in Chapter 2, you are right on track.) "Self-organization," Abraham concludes, "is where a system can create changes, including bifurcations, by *influencing its own control parameters…*" (pp. 16-17).

Anyone encountering this subject for the first time might find it bewildering, and it can be! Bewildering, but also beautiful because it leads right to this jewel: "Emergent properties of the network [i.e., system] can flow from these self-organizational features" (p. 17). This statement is metaphorically a jewel because it states the crucial relationship between self-organization and emergent properties in a system.

Abraham summarizes his discussion of this relationship by quoting J. Goldstein (1999): emergence is "the arising of novel and coherent structures, patterns and properties during the process of self-organization in complex systems" (p. 49).

Emergence in the above sense is a key concept in understanding the power and subtlety of the CDST model. I have already linked emergent properties to evolution. Furthermore, I intend to show that it is foundational in grasping the nature of mind, consciousness, intelligence—and genuine education.

A Parable Might Help

For a slight change of pace I offer a little story featuring a cast of characters whose names you are coming to know. The story is inspired by the proverb "all roads lead to Rome," which has been popularized in many ways, including as a title for both a film and a song. In the current context, I suggest that roads stand for *trajectories*, while Rome stands for a particular *attractor*. In the context of those metaphors, a *bifurcation* means a fork in the road: one fork might still lead to Rome, whereas the other fork heads off to quite another destination—another attractor.

To develop the metaphor, imagine a wandering pilgrim as a *complex system* ("a set of interacting factors") with many characteristics (variables, in that they wax and wane as time goes on). For the sake of simplicity we will focus on just two of those characteristics: a need to feel virtuous, and a need to have fun. The pilgrim permits us curious observers to follow along and measure the values of these characteristics at intervals along the journey. (Technical note: the numeric values of these two characteristics taken together constitute a *"vector."* A new character, relatively minor—sorry to spring it on you.)

Anyway, we measure and plot these vectors periodically along the pilgrim's progress, which enables us to know her/his state at each interval. Since our system is identified as a pilgrim, let's say that the need to feel virtuous would numerically exceed the need to have fun, and that the pilgrim's trajectory would tend toward Rome. But we suspect things might happen along the way that could change this tendency, and we are interested to find out what will actually happen.

Sure enough, the pilgrim begins to meet other travelers, all pursuing their own trajectories, who tell stories about another city that offers a variety of activities featuring sensory delights. These stories begin to influence the pilgrim's motivational state as revealed by our periodic measurements, and we understand that they therefore constitute a *control parameter*. Little by little, the measurements reveal that the pilgrim's need to have fun begins to exceed her/his need to feel virtuous, and the other city becomes a competing attractor. The more stories she hears about it (i.e., the higher the value of the control parameter) the more she yearns to go there, but her pilgrim conscience keeps her on the road to Rome. We can imagine that she experiences inner turmoil, which builds to such an extent that, when she arrives at a fork in the road and finds that she must make a choice about which city to head for, she pauses in complete indecision.

She has reached a point of bifurcation. Her trajectory can either continue toward Rome or veer off toward Fun City. Either choice will involve a major change in the pilgrim's inner ecology; the pilgrim/system will have evolved to a new phase, characterized by novel properties— emergent properties.

English literature provides many parables similar to this one, so we can imagine what some of those properties might be. For instance, if the pilgrim continues on toward Rome, it is likely she has a renewed confidence in her virtue, possibly tempered by deeper self-understanding engendered by the struggle with temptation that she has just experienced. If, however, Fun City proves the stronger attractor, the pilgrim might proceed with a newfound sense of freedom from constrictive conditioning about right and wrong, and begin to experience the world more expansively. In either case we see that new properties have emerged in the system that our pilgrim represents. The pilgrim has *learned*; the system has *evolved*.

No matter which fork the trajectory takes, it is important to notice that the bifurcation was preceded by considerable turbulence in the pilgrim's state space as the two attractors competed with each other, internalized by the pilgrim as conflicting emotional pulls. Such turbulence appears to be a regular feature in complex systems, signaling *the approach of a bifurcation and consequent phase shift.*

In the interest of making this metaphoric parable as instructive as possible, imagine tracking the trajectories of numerous pilgrims rather than just one. Let us say that they all began their respective journeys from many different starting points, and that each pilgrim embodies a unique combination of "need-to-feel-virtuous/need to have fun." Imagine further that they all hear the same alluring stories about Fun City—meaning that the control parameter is the same for all of them.

A plot of all these trajectories, constituting a *phase portrait* (another new character, relatively minor for purposes of this story), would likely reveal a tendency to converge toward the Rome attractor, along with all of them approaching a critical point of bifurcation as they encounter more and more turbulence-inducing tales about Fun City.

At that bifurcation point a very interesting pattern will be revealed: some of the trajectories will continue on toward Rome, while others will veer off toward the Fun City attractor. Still others might opt for neither attractor, choosing instead to wander into the wilderness, seemingly at random but actually in pursuit of some other attractor that we observers did not expect and know nothing about.

That would exemplify what Abraham, Gleick (in his 1987 classic, *Chaos)*, and many others have called a strange attractor. The apparently random trajectories responding to a strange attractor have been termed chaotic, with the proviso that the word connotes not disorder but rather *a realm of order that is subtle and hidden.* Devotees of complex systems theory thrive on discovering and mapping such previously unknown, unexpected, subtle realms of order. Every such discovery brings us to the edge of mystery, ever closer to comprehending the miracle of emergence, understanding how something never anticipated within a linear model can suddenly show up in our universe.

On that note, the parable ends.

Mathematically-derived, computer-generated three-dimensional images of this subtle realm have enabled theorists to posit much about it that will be taken up in the chapters to follow, as we explore how the CDST modeling strategy enables deeper understanding of the phenomena addressed in this book. Before proceeding, however, one more aspect of strange attractors and chaos remains to be emphasized: *sensitivity to initial conditions*. This phrase is found repeatedly in explorations of complex systems (often in the company of the phrase "far from equilibrium"). A pioneer in the study of human consciousness, Stanley Krippner (1991) stated for instance: "... in the mathematical models of chaos one encounters sensitivity to initial conditions, where even the smallest difference in the initial conditions can lead to a large difference later on within a chaotic attractor." For a school-related illustration of this axiom, imagine two children arriving at school, one without breakfast and the other sufficiently nourished (a small difference in initial conditions), but otherwise equally capable students. They both take a difficult test. The lack of nutrients in the cerebral blood supply of the undernourished student impedes her ability to think clearly, and she fails the test. The other student, unimpeded by a lack of nutrients but otherwise no better prepared, is able to recall the subject material more clearly and receives a passing grade (a significant difference in outcome).

MIT professor Edward Lorenz discovered this principle in 1960 and it became popularly known as "the Butterfly Effect—the notion that a butterfly stirring the air today in Peking can transform storm systems next month in New York." (Gleick, 1987, p. 8). It was Lorenz who plotted the trajectories shown a few pages ago in figure c—an iconic illustration of a strange attractor.

I hope that this crash course in the basics of complex dynamical systems theory—CDST—has stirred the air of the reader's understanding with a bit more vigor than might be expected from a butterfly's wings. The chapters to follow will bring these basics to bear in the pursuit of deeper insight into the most elusive and fascinating aspects of human existence: the embodied brain/mind, consciousness, intelligence, and how best to nurture them in their evolutionary march. As that insight unfolds, I further hope for a storm of positive transformation to emerge from the stirring.

References

Abraham, F. (2014). A Beginner's Guide to the Nature and Potentialities of Dynamical and Network Theory, Part 1. *Chaos and Complexity Letters.* 8 (2-3), 1-30.

Buzsáki, G. (2006). *Rhythms of the Brain.* Oxford University Press.

Giroux, H. (2016). *"We No Longer Live in a Democracy": Henry Giroux on a United States at War With Itself.* TruthOut. https://truthout.org/articles/we-no-longer-live-in-a-democracy-henry-giroux-on-a-united-states-at-war-with-itself/

Gleick, J. (1987). *Chaos: Making a New Science.* Viking Penguin.

Goldstein, J. (1999). Emergence as a Construct: History and Issues. *Complexity and Organization.* 1(1), 49-72.

Kincheloe, J. (ed.) (2005). *Classroom Teaching: An introduction.* Peter Lang.

Krippner, S. (August 15-16, 1991). Inaugural Conference of the Society for Chaos Theory in Psychology, Saybrook institute, San Francisco.

Van Regenmorte, M. H. V. (2004). *Reductionism and complexity in molecular biology.* National Library of Medicine. https://pubmed.ncbi.nlm.nih.gov/1552079

Chapter Four

Deepening Systems Understanding

The previous chapter ended by launching a discussion of chaotic (also known as strange) attractors, and of the subtle, hidden realms of order they are suspected to entail.

J. A. S. Kelso (1995), whose evidence-based theoretical breakthroughs will be discussed at some length in the current chapter, apparently has this order-in-chaos concept in mind when he says:

> What one always finds at the heart of the evolution of complex behavior are dynamic instabilities, bifurcations of different kinds…Complex systems…seem to live near these instabilities where they can express the kind of flexibility and adaptability that are fundamental to living things. (p. 22)

To grasp the above statement is to understand a very important characteristic of complex systems. *A degree of instability* is necessary to their functioning. If there is too much stability in the system, the energy to accomplish change—to adapt to changing conditions—might not be available without adversely affecting other parts of the system.

A little farther along, Kelso elaborates:

> In self-organizing systems, contents and representations emerge from the systemic tendency of open, nonequilibrium systems to form patterns…A lot of action—quite fancy, complicated behavior—can emerge from some relatively primitive arrangements given the presence of nonlinearities. (p. 34)

In a more recent publication, Kelso (2008) explores more deeply the "systemic tendency of open, nonequilibrium systems to form patterns" and the emergence of "complicated behavior" alluded to above, introducing the rubric "coordination dynamics":

> Coordination dynamics stresses…that the organism and the environment are complementary. Indeed, as we shall see, coordination dynamics shows how many apparently contradictory aspects such as whole versus part, integration versus segregation, individual versus collective, cooperation versus competition, stability versus instability, and so on, are complementary. In doing so, coordination dynamics opens up a path to reconciling contradictions, dualisms, binary oppositions, and the like in all walks of life, illuminating thereby their complementary nature… (p. 185)

In the above passage, Kelso embraces an epistemology reminiscent of Eastern wisdom traditions such as Advaita[8] Vedanta, which is innately appealing to me. I want to know as much as possible about "reconciling contradictions, dualisms, binary oppositions…in all walks of life."

Kelso elaborates further:

> New empirical and theoretical developments in the science of coordination suggest that the reason the mind fragments the world into dichotomies (and more important, how opposing tendencies may be reconciled) is deeply connected to the way the human brain works, in particular its multi- and metastable dynamics… (pp. 185-186)

His ensuing description of "metastable dynamics" corresponds to the discussion of "bifurcation" in the pilgrim parable offered earlier, moving us ever closer to using complex dynamical systems modeling for a deeper understanding of our brains. In Kelso's words:

> Metastability is an entirely new conception of brain functioning where *the individual parts of the brain exhibit tendencies to function autonomously at the same time as they exhibit tendencies for coordinated activity*… Metastability's significance lies not in the word itself but in what it means for understanding informationally coupled, self-organizing

---

8   The literal English translation of this Sanskrit word is "without duality," a concept that is re-visited in Chapter 7 in discussing Abraham Maslow's very westernized formulation of self-actualization.

dynamical systems like the brain and its complementary relation to mind. (p. 187)

Italics were added in the above quote to accentuate the essential feature of metastability in brain functioning—the existence of both autonomous functioning of its component sub-systems and coordination among them. Kelso supports this statement with a meticulous review of "empirical evidence that the structural units of the brain that support sensory, motor, and cognitive processes express themselves as oscillations with well-defined spectral properties," much of which is covered in *Anatomy of Embodied Education*. He speaks of "synaptic connections between areas" and how "phase coupling...allows groups of neurons in distant and disparate regions of the brain...(to) synchronize together" (p. 187).

He sums up his review:

> ...nonlinear coupling among oscillatory processes that possess different intrinsic frequencies is necessary to generate the broad range of behaviors observed including pattern formation, multistability, phase transitions...patterns of behavior arise as an emergent consequence of self-organized interactions among neurons and neuronal populations and this self-organization is a fundamental source of cognitive, affective, behavioral and social function. (p. 188)

Bear in mind that these assertions are based on rigorously obtained research results, and intricate mathematical modeling of those results—employing terms that are by now familiar (attractor, bifurcation, parameter, and so forth). It is beyond the scope of this chapter to describe all of this in detail, so I must be content with reporting Kelso's overall conclusions:

> Individualist tendencies for the diverse regions of the brain to express their independence coexist with coordinative tendencies to couple and cooperate as a whole. As we have seen, in the metastable brain local segregative and global integrative processes coexist as a *complementary* pair [emphasis added], not as conflicting theories. Metastability,

by reducing the strong hierarchical coupling between the parts of a complex system while allowing them to retain their individuality, leads to a looser, more secure, more flexible form of functioning that promotes the creation of information. Too much autonomy of the component parts means no chance of their coordinating and communicating together. On the other hand, too much interdependence and the system gets stuck; global flexibility is lost. (p. 196)

Kelso has thus presented highly convincing evidence that supports what Buzsáki (2006) had stated more generally in *Rhythms of the Brain* two years earlier:

If a system, for example, a neural network, can self-organize in such a way as to maintain itself near the phase transition, it can stay in this "sensitized" or metastable state until perturbed...the most basic functions accomplished by neuronal networks are pattern completion and pattern separation, functions related to the concepts of integration and differentiation. (pp. 64-65)

The terms "pattern completion and pattern separation" correspond precisely with Kelso's "global integrative" and "local segregative" processes. Together, the two processes are central to coordination dynamics. Echoing Buzsáki's statement above, Kelso (2008) explains that "...if internal or external conditions change when the system is near instability, a bifurcation or phase transition may occur..." (p. 193), meaning that the system's trajectory switches to another attractor.

Remember that this is just what happened in the parable earlier, in which the pilgrim's trajectory, which had been "integrative" with respect to the Rome attractor with all of its construed meanings (including the companionable relationships with fellow pilgrims), became more and more unstable as tales of Fun City continued to pour in. In the parable, our pilgrim's state space reached a level of instability at which a competing attractor might force the trajectory to bifurcate, to swerve toward this new attractor. If that should happen, it would demonstrate a "local segregative process" in the coordination dynamic, replacing at that point in time the "global integrative process" that had held sway up until then. The phase portrait would have changed; the system would have evolved.

We can imagine that a new integrative process would begin after the bifurcation, perhaps a more informed and adaptive one than the one replaced. Kelso (2008) describes just such a scenario, characterizing it as a kind of "reverie":

> ...thoughts come and go fluidly as the oscillatory units of the brain express both an interactive integrative dynamic and an individualistic segregative dynamic. Metastable coordination dynamics also rationalizes William James's (1890) beautiful metaphor of the stream of consciousness as the flight of a bird whose life journey consists of "perchings" (viewed here as phase gathering, integrative tendencies) and "flights" (phase scattering, segregative tendencies). Both tendencies appear to be crucial: the former to summon and create thoughts, the latter to release individual brain areas to participate in other acts of cognition, emotion, and action. (p. 202)

A living system—in this case, an embodied brain/mind—is thus portrayed in dynamic images of streams, of perchings and flights, its components alternately joining in synchrony, then separating to perform specialized tasks, then re-synchronizing to integrate whatever learning has been gained through those tasks. In this way does evolution proceed. And what affects these streams, whether for good or ill? Kelso goes on:

> ...environmental, intentional, attentional, emotional, learning, and memory processes are all capable of both stabilizing and destabilizing the coordination dynamics... The neural mechanisms of parametric stabilization by intention are beginning to be uncovered...to the extent these influences may be said to control the mind, this is the mind controlling itself. (p. 203)

Note that Kelso's list of influences on coordination dynamics begins with the environment, then shifts inward to acknowledge familiar cognitive processes: emotions, learning, memory, as well as intention and attention. The latter two—attention and intention—will be explored quite thoroughly as this book proceeds.

Before launching that exploration, though, more needs to be said about the role of environment in coordination dynamics. As might be expected, the emphasis on environmental factors in brain/mind functioning shows up throughout neuroscience literature. For example, Juarrero (2010):

> Feedback relations with the environment recalibrate the internal dynamics of complex systems to incoming signals. Doing so embeds the system in its contextual setting by effectively importing the environment into the system's very dynamical structure. (p. 2)

And Kelso reiterates the point when, in the concluding paragraph of his 2008 essay, he says, "Context matters."

The remainder of his conclusion will usher us back to the description of brain/body/mind interactions, equipped with a better understanding of the science to which he alluded. Speaking of "the coordination dynamics of thinking," Kelso (2008) says:

> ...the two "forces" that drive coordination dynamics deal fundamentally with meaningful information exchange in living things. One force is the strength of coupling between the elements; this allows information to be distributed to all participating elements and is a key to integrative, collective action. The other is the ability of individual elements to express their autonomy and thereby minimize the influence of others. Self-organization in the metastable regime is the interplay of both. This is the architecture of mind—metastable mind. (pp. 203-204)

Please keep the above statements readily available as this discussion loops back to crucial topics introduced earlier: mind, consciousness, intelligence, creativity, and how all of these pertain to learning, the project of education.

Tying it All Together

Recall that Daniel Siegel was quoted near the end of Chapter 1, saying "...So the issue is to go deeply into the brain and relationships to understand how this emergent process called mind actually is shaped by this interaction of relationships and the brain."

He had touched briefly on complex systems theory, making it clear that it is the bedrock of his view of mind. He emphasized that he was... "talking about three fundamental aspects of one reality. The one reality is energy and information flow. One aspect of that one reality is the sharing of energy and information flow, called relationship; another is the mechanism of the flow through the body (i.e., the brain)."

The fact that Siegel posits information flow as part of "the one reality" reflects Kelso's conclusions regarding "meaningful information exchange in living things" as essential to coordination dynamics. Siegel's insistence on the relationship aspect of "the one reality" furthers the congruence between Siegel's thinking and Kelso's.

Another neuroscientist, Wolf Singer (2009) of the Max Planck institute for Brain Research, has also weighed in on this issue:

> ...We have to consider more and more that the brain is a member of a socio-cultural network and that some of the phenomena that seem to be so difficult to explain in pure neuronal terms will have their explanation only when considering interactions among brains, or networks of brains... We shall have to consider the fact that our brains are the product of their embedding in a complex cultural environment... (pp. 327-328)

How These Formulations Apply Specifically to Education

The concept of embodied brains being embedded in a complex cultural environment might seem less abstract if we consider the educational system to be one such environment, and the brains embodied within that environment to be not only those being taught—the students—but those of everyone involved in the teaching process.

No one can deny the high-dimensional complexity of such a system as that, nor the consequent need of a modeling strategy subtle enough to enable us to understand the system better *and to help it become more intelligent!* Complex dynamical systems theory (CDST) appears to be a modeling strategy with the requisite subtlety and variety of methods, the best one currently available to us.

Recognition of the value of applying CDST especially to the system of education in the US and elsewhere exists in the form of a journal article in the early 1990s titled "Reconceptualizing Learning as a Dynamical System." The author, Catherine D. Ennis (1992), proposed that:

> Dynamical systems theory provides a framework for defining and examining critical components in complex, evolving environments. The theory offers rich models or metaphors to guide how we view complex ecosystems like those involved in learning. (p. 116)

In the article, Ennis laid out most of the characteristics of CDST that were known at that time—the characteristics we have just been exploring—and applied them deftly to issues of great interest in education.

For instance, she had this to say about attractors:

> Attractors act as bowls or basins where observable behaviors pool. Stable systems have deep attractor basins. Behaviors associated with the teaching-learning process are similar to objects drawn by gravity into the center of the basin. The attractor's stability controls and limits objects' range of movement, just as the stability and controlling nature of value systems mediates the acceptance or rejection of content or methods from competing perspectives.

> It takes a powerful, dynamic event to disturb the system to the extent that objects are lifted up out of the attractor basin and drawn toward an alternate basin. Similarly, it takes a major event to cause a stable value orientation to change or spontaneously reorganize into a new value perspective. (p. 120)

The article continues in this vein, exploring stability and instability, bifurcations, the interplay between attractors and constraints, and the application of all these dynamical features both to the process of learning and to the environment within which that process unfolds. Regarding the environment, the author stated:

> A third category of constraints includes social, economic, and political conditions that control or facilitate opportunities for learning. Contextual constraints often appear as multiple, contradictory, and overlapping perspectives that directly or inadvertently shape school and community policy...Political constraints act as powerful forces to modify the teaching-learning process. (p. 126)

Clearly, she intended by this to extend the features of CDST to the modeling of systems at different scales. A vital characteristic of CDST is that it *does* lend itself to this sort of extension.

Dropping back to the scale level that entails the learning process per se, Ennis offered this conclusion:

> Advocates of dynamical systems theory argue for a greater focus on the critical junctures or bifurcations of the process as attractors become unstable. The evolution of learning within multiple attractor basins occurs as we restructure knowledge and values within learner, instructional, and contextual constraints. Using our current understandings of attractors and constraints, we might shift from analyzing stability to examining change. Dynamical systems theory encourages us to focus our attention on the critical junctures in the learning process as beliefs and knowledge spontaneously reform to create order out of chaos. (p. 130)

Ennis had made an adventurous and entirely respectable early step in framing the ancient subject of human learning in terms of CDST. Within a few years of her paper's publication, Kelso's book *Dynamic Patterns: The Self-Organization of Brain and Behavior* was published. In it he would develop, refine and extend the same framework to the study of brain/mind and—very interestingly—the learning process.

Coming on the heels of the Ennis article, it is fascinating to review some of Kelso's conclusions (1995) on this topic:

> The concepts and methods developed here lead to a rather different picture of the learning process. Far from one thing changing, typically observed as improvement in a single task, we will present evidence that the *entire attractor layout* is modified and restructured, sometimes drastically, as a given task is learned. Learning doesn't just strengthen the memory trace or the synaptic connections between inputs and outputs; *it changes the whole system.* (p. 161)

Let's pause to absorb what the two passages quoted above are telling us: that the process of learning is far more complex than "improvement in a single task," or the acquisition of bits of information. Now we begin to see more clearly the relevance of CDST to the project of education. Applying the notion of attractors to the learning process enables a wider grasp of that process, a perspective that can reveal startling discoveries.

Kelso continues:

> We view learning very much as a molding or sculpting of intrinsic dynamics…We say that a behavioral pattern is learned to the extent that the intrinsic dynamics are modified in the direction of the to-be-learned pattern. (p. 163)

To make this statement more concrete, picture the "to-be-learned" pattern as a specific behavioral change, such as learning to juggle. That requires complex changes in neuro-muscular patterns, which certainly involves modification of intrinsic dynamics "in the direction of the to-be-learned pattern" (p. 163). Then, says Kelso, "once learning is achieved, the memorized pattern constitutes an attractor, a stable state of the (now modified) pattern dynamics."

This principle can be generalized to any kind of pattern that is to be learned, such as the acquisition of meaningful information, the ability to change one's breathing pattern during meditation, or the ability to modulate one's state of attention within a shifting environment. In any of these events the learned pattern becomes "an attractor, a stable state of the (now modified) pattern dynamics."

The last few lengthy passages have been included at this point for three reasons. One is to offer a glimpse of the historic onset of CDST's application to human inquiry; another is to affirm the breadth of its applicability. We have seen that this modeling strategy can be fruitfully applied to systems ranging from neural networks (a neurophysiological system), through the process of learning (a psychological system), all the way to a macroscale-level system such as a paradigm of education (a socio-cultural system).

The third and most important reason for delving into the intricacies of CDST-facilitated discoveries—especially those reported by Kelso—is that they dramatically advance our understanding of *how mind arises within the human brain, and how mind in turn relates intimately and reciprocally to the phenomenal world.*

On introspection, many readers will realize that they already have an intuitive understanding of the assertion just made: that mind relates intimately and reciprocally to the phenomenal world. The following chapter is devoted to girding up and fleshing out that intuitive understanding, using the systems tools laid out thus far.

References

Buzsáki, G. (2006). *Rhythms of the Brain.* Oxford University Press.

Ennis, C. (1992). Reconceptualizing Learning as a Dynamical System. *Journal of Curriculum and Supervision.* 7(2), 115-130.

Juarrero, A. (2010). Complex Dynamical Systems Theory. www.cognitive-edge.com.

Kelso, J. A. S. (1995). *Dynamic Patterns: The Self-Organization of Brain and Behavior.* The MIT Press.

Kelso, J. A. S. (2008). An Essay on Understanding the Mind. *Ecological Psychology.* 20(2), 180-208.

Singer, W. (2009). The Brain, a Complex Self-organizing System. *European Review.* 17(2), 321-329.

Chapter Five

Expanding the View

At this point it is appropriate to review the comprehensive list introduced in Chapter Three—a list of complexly-interrelated goals toward which this book and its companion, *Anatomy of Embodied Education,* are directed:

1. Understanding the structure, functions, and development of the embodied human brain;

2. Understanding the relationship of consciousness to the brain's structure, functions, and development;

3. Defining the relationship of consciousness to mind;

4. Exploring the roles of attention and intention in how that relationship manifests;

5. Suggesting the purpose and nature of intelligence;

6. Evoking and applying a modeling strategy subtle enough and robust enough to encompass all of the above, thus enabling humans to better understand our world and to connect with it more adaptively.

The intent is to proceed through the list, not step-by-step (that would be linear!) but rather in an interweaving fashion as required by its complex nature. Interweaving in this way will increase our flexibility in augmenting and summarizing overall progress toward meeting the goals, both singly and collectively.

Item Six

Since the final item on the list is so intrinsic to all the others, it seemed best to lead off the discussion with this explanatory paragraph about it:

The modeling strategy referred to in the final item listed above is, of course, complex dynamical systems theory, CDST. Much care has been taken in previous chapters to render this theory comprehensible to a motivated lay readership. In the following discussion, aspects of CDST will weave throughout consideration of the other items on the list, in hopes that it will help clarify and integrate them.

Items One and Two

As noted before, the goal listed as number one has been the topic of most of the chapters of *Anatomy of Embodied Education*. With that content in mind, consider what neurophilosopher Evan Thompson (2014) has written about consciousness: "In the simplest sense, consciousness is an awareness of the outside world. And this world need not be the world outside one's mammalian fur. It may also be the world outside one's cell membrane" (p. 110).

Equating consciousness with "an awareness of the outside world" is simple enough, but the simplicity begins to evaporate as he expands on what the "outside world" consists of. As he hints above, that is a matter of scale. In his very next paragraph, he states "…some level of awareness, of responsiveness owing to that awareness, is implied in all autopoietic systems." ("autopoietic" is synonymous with self-organizing.) Cells, the organs they constitute, and the bodies comprising those organs are all autopoietic systems in his view, and consciousness in some form exists at each of those levels.

Later he dives into more detail, referring to processes involved in specific states of consciousness: a) wakefulness and b) REM (Rapid Eye Movement) sleep (the sleep stage during which dreaming occurs).

Consciousness itself, Thompson says, doesn't arise from sensory inputs; it's generated within the brain by an ongoing dialogue between the cortex and the thalamus (a central structure that relays sensory and motor signals to the cerebral cortex and regulates levels of consciousness and sleep). The difference between wakefulness and REM sleep lies in the degree to which sensory and motor information can influence this thalamocortical conversation. During REM sleep, sensory inputs are kept from entering the dialogue, while motor systems are shut down (you're paralyzed except for eye movements) and attention fastens onto memories. Simply put, when sensory inputs participate in the thalamocortical dialogue generating consciousness, they constrain what we experience when we have waking perception. When sensory inputs can't participate in this dialogue in sleep, we dream. To put the idea another way, from the brain's perspective—or rather from the perspective of the thalamocortical system sustaining consciousness—wakefulness is a case of dreaming with sensorimotor constraints, and dreaming is a case of perceiving without sensorimotor constraints (Thompson, pp. 171-172).

Thompson's analysis—based on well-understood processes in the human brain—is brilliant, but in order to grasp it fully let me invoke some systems principles. The "ongoing dialogue between the cortex and the thalamus" definitely refers to feedback loops between these two parts of the brain. The "degree to which sensory and motor information can influence this thalamocortical conversation" may certainly be construed as control parameters in a dynamical system. And the use of the term "constraints," applied to sensory and motor information available to the system, should sound familiar by now (recall Ennis's discussion of constraints in learning environments).

Altogether, Thompson has provided an elegant example of systems principles applied to neuroscience—in this case, how changing parameter values (the levels of sensory and motor information) shifts the state of a system (referring here to the thalamocortical system) from one phase (dreaming consciousness) to another phase (waking consciousness). The fact that this shift occurs repeatedly—on a circadian cycle—indicates the presence of a periodic attractor.

And by the way, in addition to upholding the value of CDST as a modeling strategy this example has advanced the knowledge of what consciousness is, how it changes, and how it relates to brain structure and function. Not a bad piece of work in itself, the example and its analysis also serve as a springboard to exploring the linkage between perception and consciousness.

To begin that exploration, note that Thompson has defined consciousness as "awareness of the outside world...generated within the brain by an ongoing dialogue between the cortex and the thalamus." That awareness of the outside world comes about through acts of perception, which he characterizes as follows: "To perceive is to explore the world with your sensing and moving body. Perception creates meaning through sensorimotor exploration" (p. 186).

So not only does sensorimotor input affect the state of consciousness emerging within the "thalamocortical conversation," it also affects the *quality* of consciousness—the degree of meaning arising within it and being attached to its contents. The creation of meaning is seen to result from the active process of sensorimotor *exploration.*

Thompson makes a case that this exploratory sensing and moving is vitally important to a brain that is imbued with "a capacity for imagination—for imaging its past and future"—and that is embedded "within a sensing and moving body" (p. 187).

Surely this framing of consciousness and its determinants has enormous significance for educational practice. Think about it. How can anchoring the sensing and moving bodies of our students to desks, and limiting sensory input to narrow focus on screens, text books and lectures, be construed as educational, given what Thompson is saying here? These customary practices may indeed have a place within a wide range of teaching techniques, but how to fit them in needs to be mindfully worked out.

Items Two and Three

If consciousness is awareness emerging within a complex system by virtue of the system's capacity for self-organization—that is, awareness of the world in which that system is embedded, in which it moves around and about which it knows via sensory channels—how is consciousness distinct from "mind?" Why should two terms be needed for the same phenomenon?

Daniel Siegel was quoted in Chapter 1, proposing that "The mind is the emergent, self organizing process that regulates the movement of energy and information flow." Viewing the mind as a process means that it is active, dynamic. It is best to think of mind as a verb rather than a noun (in defiance of the tendency of the English language to reify the phenomenal world).

Remember also Siegel's insistence that mind is inseparable from embodied brain and relationship. Thus mind is an emergent process that is by definition self-organizing, a process enfolded within a physiological brain/body matrix, relational by nature, and charged with regulating the "movement of energy and information flow."

Now hold that alongside what Evan Thompson tells us about consciousness: that it is "awareness of the outside world…generated within the brain by an ongoing dialogue between the cortex and the thalamus." That "ongoing dialogue" sounds very much like mind regulating the flow of information. And as mind carries out its regulatory function, it is very likely generating consciousness in the process.

At least, that is one possible conclusion when I meld Thompson's schema together with Siegel's. I don't know whether either of them would agree with this conclusion—that mind's regulating activity is what gives rise to consciousness. I offer this inference to the ongoing neurophilosophical dialogue on the relationship between consciousness and mind.

A closer look provides support for this way of relating mind to consciousness. Farther along in his book Thompson quotes neuroscientist Roman Bauer, who said

> Life in general can be considered as an outcome of the self-organization of molecules to cells and of cells to organs and organisms, and in the same way mind and consciousness can be considered as a manifestation of the self-organization of elementary bio-electrical fields to neuro-electrical macrofields in brains.

That statement still does not distinguish between mind and consciousness, but Thomson proceeds to develop the notion of different field levels underlying progressive levels of consciousness:

> Individually, each neuron senses only its local electrochemical state; collectively, however, neurons synchronize their action potentials, both locally and across large distances, and

this temporal synchronization of an enormous number of action potentials produces the coherent and large-scale electrodynamical states of the brain that correlate with various modes of consciousness. Hence,...neural synchrony—the temporal synchronization of numerous action potentials—can be seen not only as a mechanism for cognition but also as the way the sentience of individual autopoietic neurons self-organizes into the brain-level phenomenon of consciousness. (pp. 342-343)

Thinking once again of Siegel's mind-brain-relationship model, in light of the above passage, mind could be considered the action of neurons synchronizing "both locally and across large distances," giving rise to "coherent and large-scale electrodynamical states of the brain", which in turn produces consciousness at the level of the entire brain. It is a different level of consciousness from that produced by individual neurons, which are themselves self-organizing systems. Thompson, in recognition of the difference in levels, terms individual neurons sentient rather than conscious. Expanding on that difference, he goes on:

According to this way of thinking, sentience depends fundamentally on electrochemical processes of excitable living cells while consciousness depends fundamentally on neuroelectrical processes of the brain. (p. 343)

"Electrochemical processes of excitable living cells" refers to the neuronal transmission of action potentials across synapses to the next neurons in a given network—a subject covered extensively in *Anatomy of Embodied Education*. Thompson is saying that sentience, one level of consciousness, arises within each of the hundred billion or so neurons as it contributes to the mind's work of moving energy and information around as needed. Sentience arises within each cell as a result of the feedback loops that connect it to its immediate environment.

What about the more inclusive, whole-brain level of "neuroelectric processes" that give rise to consciousness, not just "sentience"? These processes, according to Thompson, occur when

...collectively, ...neurons synchronize their action potentials, both locally and across large distances, and this temporal synchronization of an enormous number of action potentials produces the coherent and large-scale electrodynamical states of the brain that correlate with various modes of consciousness. (p. 344)

This long-range neural synchrony has also been discussed in various chapters of *Anatomy of Embodied Education*, along with the various modes of consciousness correlated with different oscillatory frequencies as measured by EEG instruments. In the context of the current chapter, long-range neural synchrony takes on a more vivid aspect as one of the biological foundations for consciousness itself.

In the context of the current chapter, long-range neural synchrony takes on a more vivid aspect as a biological basis for consciousness itself.

Note also the congruence of neural synchrony with one of the two "forces" that drive coordination dynamics according to Kelso, as discussed in the previous chapter: "One force is the strength of coupling between the elements; this allows information to be distributed to all participating elements and is a key to integrative, collective action." The "strength of coupling" translates to the degree of synchronization of neural action potentials in great number; "integrative collective action" is the action of mind in moving information—distributing information "to all participating elements." It is the "neuroelectric process" that generates consciousness throughout the entire embodied brain.

Another authoritative voice in this discussion is that of Allan Combs (2010), a veteran in the study of consciousness who, like Kelso and Thompson, proposes that consciousness is a bi-level dynamic process. And, just as Kelso has done, Combs invokes William James's classic metaphor of a bird's flights and perches. In that metaphor, Combs points out,

...the flights represent moments of transition between one thought and another, from one idea, recollection, sensation, or feeling, to another. The perchings occur when our awareness stops to rest in the presence of these ideas, recollections, sensations, or feelings. (Chapter 2, "Never at Rest")

These phases—metaphorically termed flights and perchings—are further described in the following section.

Items Four and Five

These two items remain to be examined before proceeding to the next chapter:

a. how attention and intention affect consciousness and mind;

b. the nature and purpose of intelligence.

The section just preceding this one included a discussion of one of Kelso's two forces involved in coordination dynamics—the force that enables neural synchronization throughout the embodied brain, thus giving rise to "integrative collective action." The second force proposed by Kelso, "the ability of individual elements to express their autonomy and thereby minimize the influence of others," is a necessary complement to the first force.

Kelso has shown convincingly that consciousness "depends fundamentally on neuroelectrical processes of the brain," processes that comprise the "temporal synchronization of an enormous number of action potentials" producing "the coherent and large-scale electrodynamical states of the brain that correlate with various modes of consciousness."

It seems safe to conclude, then, that consciousness (in its various states or modes) emerges from the large-scale synchronization of neuronal action potentials—which is to say that *consciousness, traditionally considered a psychological phenomenon, is inextricably bound to neural activity, a physiological process.* This apparent solution to "the mind-body problem" is made possible by the deep and subtle understanding of the embodied brain as a complex dynamical system that has come about only relatively recently. Descartes deserves no blame for leaving the world a centuries-long dualistic conundrum; he simply could not know any of what this entire book is about.

So consciousness has been accounted for as the bedrock for "integrative collective action" in the brain/body. Sounds good, doesn't it? Why, then, is a second force needed, one that enables neural components to break away from the collective and function autonomously.

The answer to that is relatively simple and obvious. Embodied brains have so many tasks to accomplish that they cannot possibly accomplish all of them as collective wholes, no matter how integrated they might be. Most of those tasks need to be delegated to autonomous, specialized neuronal sub-assemblies. That is where the second force comes in—the phase scattering, segregative tendencies. Both tendencies (referred to earlier in the picturesque terms borrowed from William James, "perchings" and "flights"), Kelso declared, "appear to be crucial: the former to summon and create thoughts, the latter to release individual brain areas to participate in other acts of cognition, emotion, and action" (Kelso, p. 22).

Among the concluding remarks in his essay, Kelso wrote:

> ...environmental, intentional, attentional, emotional, learning, and memory processes are all capable of both stabilizing and destabilizing the coordination dynamics... The neural mechanisms of parametric stabilization by intention are beginning to be uncovered... (Kelso, p. 23)

Note that, among the factors he named as operative in "both stabilizing and destabilizing the coordination dynamics," are intention and attention—two immensely crucial subjects that I will now address.

Attention

Regarding attention, the authority with whose work (briefly introduced in Chapter 10 of *Anatomy of Embodied Education*) I am most familiar is neuroscientist Les Fehmi. Tested and honed over several decades, Fehmi's model of attentional states, their dynamics, and methods for optimizing them, is the most coherent and comprehensive I have yet discovered. A condensed overview of the model is offered here, with a suggested connection to Kelso's coordination dynamics. For a fuller treatment of the model, readers are encouraged to read Fehmi's book, *The Open-Focus Brain: Harnessing the Power of Attention to Heal Mind and Body.*

In an early monograph describing his model, Fehmi (2003) began with this credo: "To realize fully our human potential is to learn to be aware of, to choose flexibly, and to implement effortlessly an expanding dynamic range of attentional styles for the optimum allocation of our resources." To the extent that education exists to serve the development of human potential, we suggest that this credo be taken very seriously, to guide and be folded into educational policy and practice.

From this opening statement Fehmi's monograph goes on to describe how "attentional styles and brain wave activity are reflected in each other," defining various attentional styles and identifying the parameters by which the entire array of attentional styles may be characterized.

Those parameters are depicted in Figure 5-1 as the opposite extremes of two perpendicular axes laid out so that they overlap, intersecting at the center of the diagram. Fehmi describes the horizontal axis as depicting the "scope of attention, which extends from a spherical and unlimited *diffuse* attention at one end to a *narrow* scope and focus of attention at the other," whereas the vertical axis "depicts proximity to the experience, which extends from *objective* or separate from the contents of the attention at the top to absorbed or *immersed* attention at the bottom.

The parameters diagrammed in Figure 5-1 suggest a four-fold classification of styles, arranged in the four quadrants formed by the intersection of the two axes. We may conclude that any act of attention experienced by an individual can be mapped onto this diagram.

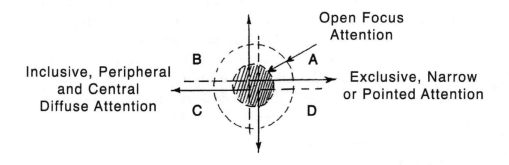

Figure 5-1. Map of Attentional Styles. From Les Fehmi, "Attention to Attention."

The table on the next page, "Styles of Attention," comparatively outlines the four classes of attentional styles in Figure 5-1, sketching the phenomenological flavor, the manifestations in the autonomic nervous system, and the signature brainwave patterns of each class.

In the viewpoint presented here, attention is a verb, not a noun, and Fehmi states very clearly that "we are always paying attention in some way...attention is something we do all the time..." In other words, attention pervades and guides every conscious experience, which makes it a supremely important feature of human existence. And yet conventional US pedagogy limits this vital subject to the most rudimentary level imaginable: the dictum "pay attention" with which we are all familiar. In the terms provided by Fehmi's model, this dictum puts forth an atttentional style characterized by extremely *narrow* scope applied to extremely *objective* content as the be-all and end-all of attentional styles.

Styles of Attention

	Diffuse-Immersed	Diffuse-Objective	Narrow-Immersed	Narrow-Objective
Example	Meditation with mind unself-conscious & body at rest. Most rapid normalization. Sleep. Most relaxed.	Panoramic view in a "symphony of sensory experience." Objective sensations hang suspended in the midst of a diffuse awareness of space. Playing in a band.	Immeresed in enjoyment, amplified by a narrow focus to intensify & savor experience. Enraptured thinker. Deep massage recipient.	Lion stalking prey. Emergency. College exam. Obsessing on work to narrow focus away from (deny) an emotional problem. Self-conscious mind & body highly toned.
Effects	Parasympathetic nervous system dominance. Low arousal. Right brain dominate. Drifting into sleep.	Relative sympathetic & para-sympathetic balance. Moderate arousal. Relative left-right hemisphere balance. Relaxed alert.	Relative sympathetic-parasympathetic balance. Moderate arousal. Left-right hemisphere balance. Alert relaxed.	Sympathetic nervous system dominance. High arousal and adrenaline. Left brain dominant. Flight or fight.
EEG	Low frequencies dominant at high amplitude. Most whole brain synchrony.	Middle frequencies dominant in amplitude. Moderate whole brain synchrony.	Middle frequencies dominant in amplitude. Moderate whole brain synchrony.	High frequencies dominant at low amplitude over-all. Least whole brain synchrony.

Figure 5-2. Styles of Attention. From Les Fehmi, "Attention to Attention."

This extremely limited approach to attention exists not only in schools; it is pervasive throughout western culture. Fehmi repeatedly points out this unfortunate delimitation, warning that "While rapid and complete attentional narrowing and objective focus is at times necessary for optimal behavior, there is, in our day, an unfortunate tendency toward overuse and consequent chronic rigidity of narrow-objective attentional processing."

It is natural to wonder how this rigid overuse of a single attentional style came about, considering the vast range of attentional styles that are available to us. Quite obviously, the potential range of such acts far exceeds the narrow-objective mode so rigidly emphasized in western culture.

We must also point out that habitual adherence to a single mode of attention directly violates the principal of requisite variety in systems. As discussed by Prof. Scott Page in lecture four in his Great Courses series *Understanding Complexity*, "robustness in a system requires that for every disturbance there must exist a response." That is essentially the principle of requisite variety. Robustness refers to the ability of a system to maintain functionality despite a disturbance (such as, for instance, a radical change in climate). Obviously, relying on only one of a great variety of possible attentional styles does not serve the need for robustness.

Fehmi has not only articulated the set of possible styles, he has also developed techniques for training individuals in adopting and applying methods for accessing a requisite variety of these attentional styles in everyday life. The rubric he has chosen for this training is "Open Focus," defined as "...the simultaneous integration of all the described styles (quadrants A, B, C, and D) of attention...This represents an integrated form of attention in which we perceive the whole field of available experience while entering upon what is relevant or most important."

This "integrated form of attention"—Open Focus—is graphically illustrated by the cross-hatched circle at the center of Fig 5-1. I propose this state of simultaneous integration of all attentional styles—and ready access to any of them depending on the need at each moment—as a potential operational definition of what has generally come to be known as "mindfulness." At the time of this writing, mindfulness practices are finding greater and greater acceptance in educational settings. It is somewhat challenging to discover a commonly accepted definition of what these practices comprise, and an operational equivalent addressing not only phenomenological aspects, but also neurophysiological concomitants, seems quite useful to me.

About the state of Open Focus, Fehmi further elaborates:

> The flexibility of attention to changing content is associated with the alternate stabilization and subsequent destabilization of various degrees of in and out of phase coherence. This mechanism is proposed as the foundation for timely ever-

changing objective knowledge, creativity, performance, and, in general, life as we know it.

Note the equivalence of Fehmi's language above to that of Thompson, quoted earlier: "coherent and large-scale electrodynamical states of the brain that correlate with various modes of consciousness…" translates to Fehmi's "in-phase coherence" (also called brainwave synchrony).

Note also the similarity of Fehmi's "alternate stabilization and subsequent destabilization of various degrees of in and out of phase coherence" to Kelso's "integrative collective action" alternating with "the ability of individual elements to express their autonomy and thereby minimize the influence of others," discussed in the previous section. It seems that Kelso and Fehmi, especially, were swimming in the same current at different times.

Controlled studies reported by Fehmi in "Attention to Attention" have demonstrated the beneficial effects of brain synchrony training (using neurofeedback), combined with Open Focus training, on grade point averages of college freshmen. Both his research and his many years of clinical experience have convinced him that "The fruits of attention practice…have staggering potential for the optimization of every phase of human life" (p. 33). "There is little doubt," he says, "that all successful learning and optimal performance involves directing appropriate styles of attention toward relevant stimuli in an effectively choreographed sequence."

I heartily agree with Fehmi's conclusion that:

> …attention training deserves the highest priority in the child-rearing process and on into and throughout adulthood and, therefore, *deserves a prominent place in our public and private school systems at every level.*

Readers, especially educators and parents, are invited to pause, take a deep breath, and give serious consideration to the quote above—particularly to the words that I italicized for emphasis.

Intention

We come now to the penultimate topic to be addressed in this chapter: intention, also known as volition or will. The subject of intention can be dealt with relatively quickly because—according to Jeffrey Schwartz, research professor of psychiatry at the UCLA School of Medicine—intention is inseparable from attention, which has just been covered rather thoroughly.

Schwartz (2002) describes the relationship between intention and attention near the end of his book *The Mind & The Brain: Neuroplasticity and the Power of Mental Force.* The final chapter is titled "Attention Must Be Paid," and in it Schwartz analyzes his long-term research involving a unique method for treating obsessive-compulsive disorder.

The four-step method that Schwartz devised involves what he terms "directed mental force," and is deeply informed by mindfulness practices. It is also, he was amazed to discover while still developing the method, completely consistent with discoveries about the inter-connected natures of will and attention reported in the writings of the great American psychologist William James in the 19th century. Schwartz found these statements by James particularly meaningful: "Volitional effort is effort of attention... Effort of attention is thus the essential phenomenon of will." On page 324 he quotes James' statements, and on the following page comes to this conclusion:

> The causal efficacy of will, James had intuited more than 100 years ago, is a higher-level manifestation of the causal efficacy of attention. To focus attention on one idea, on one possible course of action among the many bubbling up inchoate in our consciousness, is precisely what we mean by an act of volition, James was saying; volition acts through attention, which magnifies, stabilizes, clarifies, and otherwise makes predominant one thought out of many. (p. 325)

In saying this, Schwartz appears to be emphasizing only one of the many styles of attention formulated by Les Fehmi. Granted, it is the attentional style most relevant to the population he was studying. He makes this very clear later:

For the stroke victim, the OCD patient, and the depressive, intense effort is required to bring about the requisite refocusing of attention—a refocusing that will, in turn, sculpt anew the ever-changing brain. The patient generates the mental energy necessary to sustain mindfulness and so activate, strengthen, and stabilize the healthy circuitry through the exertion of willful effort. This effort generates mental force. This force, in its turn, produces plastic and enduring changes in the brain and hence the mind. Intention is made causally efficacious through attention. (p. 360)

Yet even though he focuses here on one narrow style of attention, the concluding sentence above—"intention is made causally efficacious through attention"—may well generalize to the entire spectrum of attentional styles described by Fehmi. If one's intention is to practice Open Focus and thereby gain ready access to a wide range of attentional styles, it does take effort to do that. The experiences brought about by expending that effort, the increased functionality brought about by the practice, the sense of freedom, of expanded possibilities, of course reinforce the intention.

Rewarding experiences like that are potentially available to anyone who chooses to expend the effort required. The freedom from symptoms enjoyed by OCD patients who learn Schwartz's four-step process of directed mental force, and the freedom enjoyed by Open Focus adepts—freedom from rigid, culturally induced styles of attention—are fundamentally the same freedom. Mindfulness could be said to be at the heart of each of them.

Finally, Intelligence

Considering all that has been said thus far about the multiplex brain, mind, consciousness, attention and intention, how can we best understand the nature and purpose of intelligence?

I propose that intelligence be considered nothing less than an overarching concept comprising all of the above—embodied brain, mind, and consciousness, all abetted by attention and intention. In

Chapter 2 of *Anatomy of Embodied Education,* Tim Burns and I offered a definition of intelligence as "the capacity to respond to new situations, adapt to changing conditions, and learn from experience." In this current book a model has been explored that accounts in detail for that capacity to respond adaptively. This exploration has led me to expand our initial definition, resulting in the proposal at the beginning of this paragraph.

Furthermore, I suggest now that the *purpose* of human intelligence, individually and collectively, is to keep our species, homo sapiens, on track and viable in a world that is ever-changing. That means especially guiding us humans to preserve the world on which our well-being depends, understanding *to the core* our interdependence with each other and with our living planet.

The *purpose* of human intelligence, individually and collectively, is to keep our species, homo sapiens, on track and viable in a world that is ever-changing.

It seems obvious to me that the purpose of *education*, simply stated, is to foster and refine human intelligence as conceived above.

References

Combs, A. (2010). *Consciousness Explained Better: Towards an Integral Understanding of the Multifaceted Nature of Consciousness.* Paragon House.

Fehmi, L. (2003). Attention to Attention. *Applied Neurophysiology and EEG Biofeedback*, J. Kamiya. (Ed.). Future Health.

Fehmi, L. and Robbins, J. (2008). *The Open-Focus Brain: Harnessing the Power of Attention to Heal Mind and Body.* Penguin Random House.

Schwartz, J. & Begley, S. (2002). *The Mind & The Brain: Neuroplasticity and the Power of Mental Force.* HarperCollins.

Thompson, E. (2014). *Waking, Dreaming, Being: Self and Consciousness in Neuroscience, Meditation, and Philosophy.* Columbia University Press.

Chapter Six

The Embodied Brain and Emergent Mind

"The role of the clinician, teacher, or parent is to facilitate the discovery of this potential for integration inherent in each of us. This is the way we awaken the mind to create health in the person's life."

–Daniel Siegel, MD, *Pocket Guide to Interpersonal Neurobiology: An Integrative Handbook of the Mind*

In *Anatomy of Embodied Education,* Tim Burns and I addressed the anatomy and physiology—both macro and micro—of the human brain, its development from conception through childhood and adolescence, the effects on the developing brain of environment, nutrition, genetic and epigenetic factors, relationships, and stress. We discussed chronobiological issues and their relationship to the intrinsic rhythmic fluctuations within the embodied brain, and explored the concept of integrating the innate intelligences of the brain, body, and heart.

Scattered throughout the abundant information about the embodied brain presented in that accompanying book, hints about how this information might be applied to the processes involved in developing the brain—i.e., education and parenting—may also be found. The foundation established in that book is complete enough to assist here in developing more completely and systematically the issues of its application. Specifically, I will explore how the accumulated information about the embodied brains of humans—how they are formed, how they work, and how they can be most effectively supported and developed—might help refine the paradigm and purposes of education in the Western Hemisphere.

The epigraph that begins this chapter focuses on a challenge that I intend to address seriously, both in this chapter and the one following. Daniel Siegel's extensive research and writing about the relationship of the mind not only to the embodied brain, but also to the interpersonal environment within which the brain is embedded, has been introduced in previous chapters. The current chapter reviews his extraordinary work, places it in a historical context, and examines more deeply its role in bringing CDST to bear in evolving the paradigm of education in the western world.

What is the Mind?

Discussion of this question, along with related issues such as the nature of consciousness, took up much of the previous chapter. These topics have occupied philosophers and psychologists for centuries, and they are certainly large enough in scope to occupy us for a while longer as this chapter unfolds.

The 17th century philosopher Renée Descartes famously separated mind from body, a dualistic conundrum that has only recently begun to be resolved by systems theory. In the late 19th century, William James, widely revered as "the father of modern psychology," wrote at length about mind and consciousness in his laudable attempt to resolve the dualistic burden inflicted by Descartes.

But it was not until Gregory Bateson came along in the 1970s and, equipped with knowledge of a new branch of science called cybernetics, began to make headway in unifying mind not only with body, but with nature in general (Bateson, 1972, 1979). His was a revolutionary step forward, which has been further refined as cybernetics morphed into General Systems Theory, which in turn helped give rise to theories involving complex systems and an understanding of the dynamics of chaos.

While these theoretical breakthroughs were occurring, neuroscience was also developing exponentially. Siegel's work draws from both currents, and synthesizes them in as elegant and sound a way as any I have seen. Here I will attempt to describe Siegel's ideas about the mind succinctly, deriving from them a coherent basis for the comprehensive developmental model to be presented and discussed in the final chapter.

Interpersonal Neurobiology

The heading above is the term used by Dr. Daniel Siegel to designate his body of work, which is detailed in his book *Pocket Guide to Interpersonal Neurobiology: An Integrative Handbook of the Mind*. While deliberately and interestingly presented in a unique, non-linear way, the *Pocket Guide* follows a practice I have noted in other books Siegel has written, weaving together an intricate, many-faceted theoretical structure with passages about specific regions of the human central nervous system—the cerebrum, including its somatic interconnections—that pertain to aspects of the theory being presented.

From my standpoint, Siegel's complex interweaving of philosophic exploration with neuroscientific understanding is ideal for the purposes of the book you are now reading. Just as Chapter Two of *Anatomy of Embodied Education* featured MacLean's pioneering Triune Brain theory, this current chapter is primarily about Siegel's interpersonal neurobiology (IPNB) concepts, which will guide us into the concluding chapter that follows.

Again, What is the Mind?

The first task before us is to extricate our inquiry from the rarely-questioned tendency of the English language to reify the world of phenomena—to make "things" out of what very often turn out to be processes. That is just what happens when we speak of "the mind," as if it were some*thing* we can set apart and define in an objective way. Siegel takes this vexing tendency by the horns very early in the *Pocket Guide,* emphasizing the process aspects of mind, making it more a verb than a noun. He reports his very first working definition, from which the entire field of interpersonal neurobiology eventually emerged:

> A core aspect of the mind can be defined as an embodied and relational process that regulates the flow of energy and information. (Siegel, 2012, p. xxvi)

He had introduced the need for such a definition two pages earlier by saying something very relevant to the purpose of this book:

> …as educators interested in developing a strong and resilient mind in our students, not defining the term"mind" leaves

us in the dark. As parents focusing our efforts on helping our children develop healthy and flexible minds, not having even a working definition of the mind as a starting place limits us. (p. xxiii)

The above passages appear in the introduction to Siegel's *Pocket Guide*, setting the stage for the first chapter, titled simply "Mind." That chapter begins by acknowledging a quality that is already familiar to many of us who have done some thinking about the mind. "Mind," he says, "relates to our inner subjective experience and the process of being conscious or aware" (p. 1).

Presumably, consciousness and awareness are topics of great interest to readers of this book, and it would be useful to know the relationship between them, and to mind in general. But stating the topics like this puts us back on the path toward reification. Paying more attention to his actual wording, we see that Siegel is careful to state these phenomena as "the *process* of being conscious or aware."

The distinction between mind as a thing and mind as a process is crucial to the goals of this chapter. To clarify further the importance of that distinction, consider this statement by Arthur Deikman (1973), who was professor of psychiatry at UCSF and a scholar/practitioner of Eastern wisdom traditions: "Activity, change, process—these are the substance of our bodies, of our world, of the universe. Gradients, not boundaries, determine form" (p. 318). These words were published nearly half a century ago, and summarize teachings from a long time before that. We educators and parents would do well to take them seriously— to recognize and correct excessive reification. Let the perceived world no longer consist of "things," but rather *processes*.

What we have been taught to consider "things" are actually fluctuating phenomena. Perceiving them like that opens our perception from frozen to flowing, and leads to greater possibilities for successful adaptation.

Gregory Bateson, mentioned earlier as a major architect of this way of addressing issues related to mind-as-process, was a contemporary of Deikman. The two of them had very similar ideas, as demonstrated by this statement near the end of Bateson's "Last Lecture" in 1979: "...I have offered you the idea that the viewing of the world in terms of things is a distortion supported by language, and that the correct view of the world is in terms of...dynamic relations..." (Bateson, 1991, pp. 7-13). A blunt and concise statement indeed, summarizing a major portion of his life's work.

Both Deikman and Bateson recognized the importance of systems thinking in the study of human experience, and both contributed greatly to the understanding that has developed about systems. In these towering figures we find two major precursors of Daniel Siegel's interpersonal neurobiology, to which we now return.

In describing IPNB Siegel makes copious use of systems terminology that has already been introduced. It might be helpful to review quickly some basic characteristics and definitions. A *system* can be simply defined as *a collection of components that affect each other in various ways*. Because the components affect each other, they are said to be interconnected. In *complex systems such as the embodied human brain*, the interconnections are dynamic, that is they're constantly shifting, changing. That dynamic quality applies to all complex systems at whatever level, be it an anatomical organ system, sociocultural systems such as education and the economy, or the ecosystem of planet Earth. And that is what underlies the need discussed earlier to think in terms of processes rather than "things."

> A system can be simply defined as *a collection of components that affect each other in various ways*. Because the components affect each other, they are said to be interconnnected.

Dynamism of course implies energy, and energy has a starring role in Siegel's ideas about mind. In the glossary of the *Pocket Guide* he gives the simplest possible definition of energy, calling it "a term from physics that means the capacity to do something" (Siegel, 2012, p. A1-29). He adds that "energy comes in various forms such as kinetic, thermal... electrical, and chemical" (ibid, pp. A1-29 - A1-30). He also reminds us that "the nervous system functions by way of the flow of electrochemical energy" (ibid, p. A1-30).

Energy is particularly important to humans when it becomes patterned in such a way that it carries meaning, thereby becoming *information,* which also plays a starring role in IPNB. Yet another starring role is occupied by *relationship*, defined generally as "patterns of interaction between two or more entities." Note the recurrence of the word "patterns" in the definitions of both information and relationship.

It is worth mentioning that Bateson too had placed major emphasis on this term, such as in his discussion of "the pattern which connects" in the introduction to his book *Mind and Nature: A Necessary Unity*: "...the right way to begin to think about the pattern which connects is to think of it as...a dance of interacting parts..." (Bateson, 1979, p. 13). Earlier he had spoken of a pattern of patterns—a meta-pattern—as being the pattern that connects. Here he is taking pains to point out that the pattern that connects is dynamic, not static (a *dance* of parts *interacting*).

Remember this notion of dynamic patterns as we return to Siegel's models pertaining to mind. It will become apparent that he has latched on to a potent, durable concept, and brought it to a level that accommodates a great deal of knowledge gained since the time of Bateson and Deikman.

Energy, Information, and Relationship

Siegel proposes the image of a triangle as a way of visualizing how energy and information fit together, and how their dynamic interactions, expressed through the brain/body and shared in relationship to the environment, bring about the mind. He calls it a "triangle of well-being," describing it as a "three-pointed figure that is a metaphor for the idea that mind, brain, and relationships are each part of one whole" (Siegel, 2012, p. 4-1).

With the triangle metaphor in hand, and keeping dynamic energy and information in the picture, it is possible to expand and refine his initial definition of mind quoted a few pages back:

> We can envision the triangle as being a metaphoric map, a visual image that signifies one reality with three

interdependent facets. The triangle represents the process by which energy and information flow and how the flow changes across time. Relationships are the sharing of that flow. The brain is the term for the extended nervous system distributed throughout the whole body and serving as the embodied mechanism of that flow. (Siegel, 2012, p. 4-1)

And where does mind fit in this picture? Imagine mind at the apex of this equilateral triangle. It is, Siegel proposes, "…an emergent process that arises from the *system* of energy and information flow within and between people." That emergent process entails "something called self-organization, an emergent process that regulates that from which it arises." Put another way, "…we see mind as arising from the movement of energy in the system composed of the body and of relationships… mind, among other attributes, is in part the regulatory process that shapes energy and information flow within and between people" (pp. 4-1 - 4-2).

Now it should be more apparent why mind needs to be regarded as a verb rather than a noun, considering phrases such as "emergent *process*," "regulatory *process*," and "*shapes* energy and information *flow*." The term "emergent process" (a variant of emergent property), you will recall, was explored in the earlier chapter on CDST, Complex Dynamical Systems Theory.

A few pages back I stated the impossibility of describing interpersonal neurobiology (IPNB) without bringing in terms from systems theories, and made particular mention of *complex* systems. In his exploration of what constitutes mind, Siegel specifically acknowledges the contribution of complex systems theory to his own formulations, and I ask you now to refresh your grasp of these two fundamental aspects of complex systems: self-organization and emergent properties. (These phenomena are also central to the developmental model explored in the chapter to follow, and to the generally evolutionary themes found throughout this book.)

Especially for readers with little or no prior exposure to systems theory, I offer the following few paragraphs to reprise briefly the contents of Chapters 2 (general systems theory) and 3 (complex dynamical systems theory), thus enabling you to sail confidently through the remainder of the current chapter and land securely on the shore of Chapter 7.

Systems Principles: A Brief Synopsis

Basic Definition

Recall our basic definition from three pages ago: *a system is a collection of components that affect each other in various ways.* The more components a system comprises, and the more ways these components interact with each other, the more complex the system is. And of course, as has been repeatedly emphasized, the human brain is supremely complex (in this chapter we are expanding our view of the brain to include its embodiment and its relationship to mind and the environment, especially the interpersonal environment).

It is fair to characterize the attempt to understand how the components of such a complex system interact with each other as challenging, to say the least. Complex systems theory provides a major tool in meeting that challenge.

Self-Regulation Backed-Up/Modified by Self-Organization

To have much hope of understanding a complex system, it is necessary to adopt a mindset of system-plus-environment, for no such system exists in isolation. Besides being a central feature of IPNB, this principle clarifies much of what happens within a system. A system's evolution largely consists of adapting to conditions—also known as parameters—existing in its environment. Its boundaries must be open to energy and information provided by the environment; it adapts by sensing and responding appropriately to that energy and information. That general operation is known in systems terms as self-regulation, and we have already described how self-regulation proceeds by means of feedback pathways.

Feedback in Systems

Deviation-reducing feedback loops help a system remain viable within its environment by sensing when it is deviating from an optimal response to environmental conditions, then altering its functioning accordingly. An electro-mechanical example of such a feedback loop is a common thermostat designed to activate and deactivate a heat source in order to keep ambient air temperature within a certain range. Mammalian bodies contain more complex feedback loops, as described in Chapter 3, that serve the same purpose.

In this way, systems self-regulate within existing environmental conditions or parameters. When those parameters (climate conditions, for example) change, as they inevitably do, systems survive only by shifting to a higher level of complexity to accommodate the change in parameters. A shift like this is known in general systems terminology as self-organization—an essential aspect of what we commonly term "learning."

Only open systems with sufficient complexity undergo a shift like this—an evolutionary advancement. The shift is not imposed from outside the system, but rather emerges as a creative response to changes in the environment that render the system's previous responses ineffective.

For example, consider the evolutionary changes that must have occurred in prehistoric humans when challenged by environmental changes that led to the Ice Age. They would have had to make cultural adjustments such as dressing more warmly, changing foraging patterns, learning to make and control fire. Adjustments like this would have required internal changes as well, changes in body and changes in brain.

Emergent Properties/Processes

Somewhat paradoxically, when those adaptive changes occur—when a system shifts to a higher level of complexity, as described above—those very changes influence what goes on within the system that gave rise to them. That is what Siegel meant when he referred to "an emergent process that regulates that from which it arises." Complex systems evolve to generate novel and largely unpredictable characteristics that enable them to survive and thrive in novel environments. That is what emergent properties are, and Siegel tells us that is what mind is:

> The self-organizing processes emerging from energy and information flow in our bodies and in our relationships certainly give rise to our mental activities, which channel

that flow and function as important regulatory aspects of our mental lives. (Siegel, 2012, p. 1-9)

Back to the Triangle

Having looked more deeply into the meaning of the terms self-organization and emergent properties, we are better equipped to absorb the full import of "the triangle of well-being." Siegel (2012) says:

> And so here is our triangle in brief: regulation (mind) entails the monitoring and modifying of the flow of energy and information. Sharing (relationships) is the exchange of energy and information between two or more people. The mechanism (brain) is the structural means through which the energy and information flow occurs within the body. (p. 4-2)

Mind

Brain/body   Relationships

But why call it a "triangle of well-being?" The answer to this question brings us to the brink of fully grasping the applicability of IPNB to the purposes of this book. Having established that mind, brain and relationships are what he terms "the three primes of experience—the irreducible aspects of the system" (p. 4-3), he goes on to say (on the same page):

> When we move from a basic triangle of human experience to a triangle of well-being, we have entered the realm of addressing the question of what are a *healthy* mind, a healthy brain, and healthy relationships. From an interpersonal neurobiology perspective, *integration* is the definition of good health.

"Integration," Siegel tells us, "is the linkage of differentiated elements. A healthy mind, a healthy brain, and healthy relationships emerge from integration" (p. 4-3).

That deceptively simple phrase, "the linkage [integration] of differentiated elements," introduces the two fundamental systemic processes that underlie IPNB: integration and differentiation. We are to understand that health entails a delicate dance involving both of these processes. It appears that the entirety of IPNB consists of exploring and applying that dance to virtually every facet of human existence.

Our western world certainly knows about differentiation; we are driven, it seems, to separate, categorize, polarize, and even fragment our world—making it a world of *things*. Differentiation is a given for us.

Siegel believes however that "we each have a natural push toward integration," asserting "each individual's innate potential to 'heal' and become 'whole' by releasing the blocked capacity of that person to integrate the brain and relationships" (p. 4-6).

"The role of the clinician, teacher, or parent," he tells us, "is to facilitate the discovery of this potential for integration inherent in each of us. This is the way we awaken the mind to create health in the person's life" (p. 4-6).

The remainder of his *Pocket Guide* rolls out a plethora of inter-related aspects of the dance of differentiation and integration, along with descriptions of many methods for optimizing integration. Among those aspects, one of the most crucial is attention. Learning to deploy attention skillfully is one of the most effective methods for optimizing integration.

Siegel devotes extensive discussion to the topic of attention throughout the *Pocket Guide*. That and awareness are the two subjects referred to the most in the book's index/glossary. One of the earliest references is on pages 3-5: "The process of using attention to change the activity of the brain—and therefore ultimately its very architecture—is a part of the larger process by which experience changes neural structure."

Throughout his book Siegel expands on this fundamental aspect of *attention*—that it is instrumental in changing the activity of the brain.

He defines it as "the process that shapes the direction of the flow of energy and information" (p. 7-1). Attention thus has a cardinal role in the functioning of the brain and in the makeup of mind overall.

In this book's previous chapter, I place a great deal of emphasis on the model of attention created by Lester Fehmi. From that model, let us recall particularly the most comprehensive of the styles of attention articulated by Fehmi, the one he termed "Open Focus." This phrase denotes a kind of attention, a state of awareness, a state of *being* that may well be considered "the jewel in the center of the lotus."

Open Focus is an optimally flexible state of attention. As its position in the center of Figure 5-1 suggests, it is a state in which any of the myriad states or styles of attention available in attention space can be readily accessed, to meet the requirements of whatever situation might arise in a person's life.

Furthermore, it is a state that can be taught, and has been taught to thousands of people. Fehmi has created a teaching method (Fehmi, 2003, p. 17) that he calls "Open Focus Training," which consists of a wide variety of guided practices available as audio recordings that the trainee can listen to at home. Regularly repeated over a period of time this meditation-like practice can, thanks to neuroplasticity, make the open focus state of attention the trainee's default attentional mode, replacing the often problematic states she or he has learned unconsciously over a lifetime and uses automatically, with little awareness.

The practice is often augmented with a form of neurofeedback training that Fehmi has designed, which increases synchrony throughout the trainee's brain. This "whole-brain synchrony" is the definitive EEG pattern for the open focus style or state of attention, and brings with it a wide range of benefits for the trainee (Fehmi, 2008).

From Synchrony to Integration

Fehmi's use of the term synchrony places an emphasis on a specific physiologic phenomenon: neural assemblies firing in phase with each other. Siegel uses a more general term, integration, to characterize healthy functioning at all levels—including neural. He says for instance:

"a teacher can implement a curriculum…that supports whole-brain integration" (Siegel, 2012, p. 9-4). By whole-brain integration, he means the harmonious or synchronous functioning of all parts of the cerebrum. Thus Fehmi and Siegel share common ground.

Yet Siegel applies the term integration over a range that includes not only the embodied brain, but all aspects of the triangle of well-being. In the following passage he refers to the brain, then goes on to clarify and extend what he means by relationships:

> Applying the triangle in everyday life enables us to see how our minds emerge not only from neural mechanisms but also from relationships we have with other people and with our planet…[enabling] an expanded sense of identity that goes beyond the boundary of our skin, beyond a definition of "self" that is limited to just our bodily encasement. (Siegel, 2012, p. 4-6)

The possibilities inherent in "applying the triangle" to the project of rearing and educating young people seem boundless and profoundly exciting. Surely, infusing integrative practices such as mindfulness and Open Focus Training into parenting, classroom procedures and curriculum design would help the embodied brains, minds and relationships of future generations to evolve as needed to meet the unimaginable challenges that loom before us.

In the concluding chapter just ahead I delve even more deeply into the need for such practices, viewed from a comprehensive developmental framework that not only encompasses the early stages of human life, but reaches over the entire life span.

References

Bateson, G. (1972). *Steps to an Ecology of Mind.* Ballantine Books.

Bateson, G. (1979). *Mind and Nature: A Necessary Unity.* Bantam Books.

Deikman, A. (1973). The Meaning of Everything. in R.E. Ornstein (Ed.), *The Nature of Human Consciousness.* W.H. Freeman.

Fehmi, L. (2003). Attention to Attention. in J. Kamiya (Ed,). *Applied Neurophysiology and EEG Biofeedback.* Future Health Inc.

Fehmi, L. and Robbins, J. (2008). *The Open-Focus Brain: Harnessing the Power of Attention to Heal Mind and Body.* Penguin Random House.

Siegel, D. (2010). *The Mindful Therapist.* W. W. Norton & Company.

Siegel, D. (2012). *Pocket Guide to Interpersonal Neurobiology: An Integrative Handbook of the Mind.* Norton.

Chapter Seven

A Framework for Optimal Human Learning Throughout Lifelong Brain Development

To launch this chapter, I will reiterate a statement introduced in Chapter 2 of *Anatomy of Embodied Education*: "Intelligence is the capacity to respond to new situations, adapt to changing conditions, and learn from experience." The challenges we all face in today's world—accelerating environmental degradation, alarming disappearance of species, potentially catastrophic global heating, growing economic and social chaos—can be successfully addressed only when enough of us develop the requisite level of intelligence.

Now consider the above assertion alongside what the iconic 20th Century genius Albert Einstein has often been quoted as saying: "problems cannot be solved using the same level of thinking that created them." The implication of these statements is unmistakable: humanity must evolve its level of thinking—its intelligence—to successfully address the problems exemplified above.

Clearly, educational policies and practices have vital roles to play in this evolutionary imperative. The intention of this final chapter is to pull together cardinal aspects of the abundant information presented in previous chapters, and weave them into a model that suggests how education *in general* might make beneficial use of all this information.

I leave it largely to the practitioners themselves—the teachers, the administrators, the curriculum development specialists—to create specific environments and practices, based on the general suggestions, that best suit the specific populations that the practitioners serve. My role is to craft a model, a creative synthesis of well-founded ideas about human development on one hand, and discoveries from contemporary neuroscience on the other, that will serve as a springboard from which motivated readers can find the impetus needed to apply the model to their unique situations.

This model has two primary purposes:

1. to organize schematically the knowledge it encompasses, so that readers' understanding of that knowledge may become clearer and deeper;

2. to build a solid rationale for applying that knowledge to optimal child-rearing and teaching.

Some Notable Predecessors

A rich foundation for examining the life-long progression of human capabilities has already been laid down by Jean Piaget (1952), Erik Erikson (1963), Abraham Maslow (1971), Lawrence Kohlberg (1984), and Rudolf Steiner (1996). Tim Burns and I acknowledge the contributions of these guiding lights. Over the course of his career as a consultant and presenter at educators' conferences around the world, Burns has built the model described in *Anatomy of Embodied Education* on their foundational work. With his blessing, I intend to meld the classical developmental concepts of these pioneers with the neuroscience material and consciousness theory that permeate this book.

An Initial Look at the Developmental Framework

The illustration on the following pages (pp. 92-93) depicts a visual map of the developmental model that is the focus of this chapter. A considerable amount of text is offered to describe, explain and explore this complex graphic. Taken together, the chart and the explanatory text comprise the model toward which this entire book has been progressing.

I have already mentioned the earlier version of the model featured in our earlier book. In this current book, with the superb assistance of graphic designer Jessica Phillips, I have re-crafted the visual depiction of the model. The current presentation has been prompted by my ongoing exploration of puzzles entailed by consciousness, mind/body relationships, complexity and emergence—puzzles that were beyond the scope of the earlier book to address.

So, keeping this introduction in mind, let's move forward with the model.

The chart displayed on the following two pages incorporates several complexly-related dimensions of human development from early childhood onward. The top section lays out essential aspects of the contributions of each exemplary predecessor named on the previous page.

Notice that beneath each name, in smaller print, appears a single word denoting a primary human characteristic on which that particular researcher tended to focus during his career (e.g. cognition for Piaget, virtues for Erikson, etc.). Then, in each cell to the right of the researcher's name, appears a brief summary of the issues observed by that researcher to affect their subjects at each stage of development.

What is meant by "stage of development" in the current context? Looking at the first row of that top section, headed "Brain Stage," we see five stages of brain development (as discussed extensively in Chapter 2 of *Anatomy of Embodied Education*) shown in that row. Starting at the left with the brain stem (i.e. the medulla oblongata and the pons, picturesquely called by MacLean "the reptilian brain") the developmental stages are shown in chronological order tagged with the approximate age-range corresponding to each stage. (Notice the thick black arrow running along between the top and bottom sections; it has those age ranges embedded in it.)

The second row tabulates the level or domain of intelligence associated with each level of brain development shown in the first row.

Each of the remaining five rows in the top section is labeled by the name of one the five theorists designated earlier as "notable predecessors." Beneath each name is a one-word characterization of that theorist's focal interest. The five cells to the right of each name encapsulate how that theorist characterized the stage of overall development corresponding to the brain development stages named in row one (Brain stem/cerebellum, Limbic system, etc.).

Framework for Optimal Human Learning and Development

Brain Stage	Reptilian Brain stem/cerebellum	Paleo-mammalian Limbic system	Neo-mammalian Posterior cortex
Intelligence Domain	Body Self preservation	Social-Emotional Relationship	Thought Concrete & problem solving
Piaget Cognitive	Sensory-motor	Pre-operational	Concrete operations
Erikson Virtues	Hope: Trust v. mistrust	Will: Autonomy vs. Shame & Doubt	Purpose: Initiative vs. Guilt, Competence: Industry vs. Inferiority
Maslow Needs	Survival & Safety oriantation	Love & Affection orientation	Belonging orientation Esteem orientation
Kohlberg Moral	NA	Pre-conventional: punishment & obedience	Conventional: conformity, authority, social-order maintenance
Steiner Spiritual	Goodness orientation	Beauty orientation	Truth orientation

	Birth-2 yrs	2-6/7 yrs	6/7-Puberty
LEVELS of **EMERGENT INTELLIGENCE** Express	**Sensory-Motor Integration** Sensation, perception, initial exploration and manipulation	**Symbolic-Representational** Mental images, pictures, words, language and stories	**Social-Emotional** Relating, feeling, rule-making, role-taking and play
NATURE'S PLAN for optimal human development Supported by	**Bond** with and attach to a consistent, nurturing care provider	**Overcome** obstacles to development	**Develop** the imagination and aquire emotional-relational fluency
GUIDING STRATEGIES from a perspective of optimal development	Provide a caring and supportive environment that maximizes love and safe limits while minimizing harmful stressors	Support sensory integration through movement, play, and imagination as the foundation for learning and development	Attend to emotional-relational development as the key to learning and successful living

Neo-mammalian Frontal cortex	Brain - Heart Integration	Beyond current knowledge
Thought Abstraction & meaning making	**Heart** Wisdom & compassion	Developmental stages only recently coming under scientific exploration. For instance, neural correlates of: a) interpersonal synthesis b) identification with natural world c) unitive consciousness
Formal operations	Post-formal operations	
Fidelity: Identity vs. Role Love: Intimacy vs. Isolation	Care/Wisdom Generativity vs. Stagnation/Integrity vs. Despair	
Esteem orientation: confidence, achievement, respect for & by others	Self-actualization morality, creativity, acceptance	
Conventional: Social contract to Post-coventional	Post-conventional: principled conscience universal ethic	
Soul orientation	Spiritual orientation	

Wise Being

Teen-Adult   Mature Adult

			Transpersonal-Transrational The awakened heart: universal, boundless, metaphorical, noetic
Concrete-Creative Multi-sensory manipulation, experimentation, building and creating	**Abstract-Conceptual** Logic and reason, analysis, hypothesizing, meta-cognition, possibility thinking, idealism	**Global-Systemic** Insight, intuition, integration, meta-meaning and altruism	
			Cultivate and integrate wisdom and compassion
Discover and express creative talents, gifts and multiple intelligences	**Achieve** a sense of coherence, relevance, significance and meaning	**Connect** with a power beyond the self	
Ensure full use of the arts and science as central to the learning process, with ample opportunity for creative exploration and expression	Use authentic tasks that call forth problem solving, critical thinking, idealism and active construction of meaning	Provide opportunities for connecting with a larger sense of purpose and place	Introduce practices that awaken the heart, clarify the mind, and enliven the spirit

Looking now at the lower section of the chart, beneath the arrow depicting approximate chronological categories of development, you will see three rows that are conceptually related to the rows above the arrow, but that have a subtly different purpose. We are about to proceed from the descriptive aspects of understanding stages or phases of human development to what we might refer to as the prescriptive aspects.

As to the descriptive aspects, Tim Burns has done thorough scholarly work abstracting the major features of the developmental findings of several historic titans in the study of human development, then schematically analyzing those findings within a framework that includes neurophysiological milestones about which those pioneers (along with just about everybody else) had very little knowledge. That schematic analysis is primarily what we see in the upper section of the chart being presented.

Now, shifting attention to the lower section of the chart, we find a melding of the stage-related descriptive scheme shown at the top with an overview of how each stage of development is innately tied to: a) growing, emerging levels and expressions of intelligence throughout an optimal human lifetime; b) the underlying natural pattern/plan/purpose for each of those stages; and c) specific guidelines that parents—and later, professional educators—are encouraged to employ in order to optimize passage through each developmental stage. These "Guiding Strategies" are extensively discussed in our original book.

The change in form of the model's graphic presentation in this book compared to the earlier one necessitates some additional commentary from me. You the reader will note that the diagram comprising the lower section contains seven columns instead of the five columns up above. That is because the "Brain Stages" (from reptilian through brain-heart integration) and "Intelligence Domains" (from body intelligence through heart intelligence) extend only as far as the 20th century theorists could reach in their time.

From our current perspective though, we can see beyond those limits, and even reasonably hypothesize beyond what we can actually observe. I attempt to present this visually by showing the arrow—symbolizing the progression of phases of human development—beginning to curve upward toward advanced levels of the possible human.

That same speculative spirit is indicated on the upper part of the Framework. There, the Brain Stage continuum (top row) extends beyond Brain-Heart Integration (already supported by progressive neuroscience described, for instance, in Chapter 5 and later on in the current chapter), and toward what I refer to on this graph as "Beyond current knowledge."

Likewise, the Intelligence Domain continuum—shown on the row just beneath the Brain Stage continuum—ends on this graph with the Heart domain (sub-head "Wisdom & compassion"). In the shaded area just beyond that, beneath my extended Brain Stage heading "Beyond current knowledge," I have added the label "Developmental stages only recently coming under scientific exploration."

To summarize the matter quite simply, our predecessors explored these developmental phenomena as far as they were able to, given the concepts and research methods available to them in their time. In the current century we are witnessing a resurgence of fascination with human development comparable to that evidenced by our exemplary predecessors. What distinguishes the current resurgence is a combination of advanced research technology and expanded conceptual frameworks.

To re-state what I hope is already obvious, this book is primarily about that conceptual expansion. The fundamentals of human development summarized in the lower half of this chart, and the "Guiding Strategies" for optimizing human development from early childhood through mature adulthood presented here are still valid. It is vitally important that parents and educators understand these fundamentals and embrace the guiding strategies congruent with them. Beyond that, I am also inviting readers to embrace even higher reaches of potential development, the nature of which is suggested in columns six and seven in the lower right-hand part of the Framework.

The remainder of this concluding chapter will discuss and expand on that invitation. I assume that anyone who has read this far in the book is capable of imagining levels of emergent intelligence beyond the Abstract-Conceptual attainments of a mature, well-educated adult, having successfully navigated the previous stages. The stages leading up to this capability are discussed in Appendix A. Please feel free to view that discussion or any part of it whenever you like. Many readers may want—as would be my tendency—to peek at the crescendo of the *finale* before filling in the passages leading up to it. If that is your tendency, what follows is where you would begin your peek.

Levels Six (*Global-Systemic*) and Seven (*Transpersonal-Transrational*)

By the time any adequately-prepared human reaches these final two levels, that person will no longer be a student in a school system as we commonly understand that term. It is to be hoped, though, that such a person would be among many teachers and other leaders of school systems throughout the world whose emergent intelligence has prepared them to serve as *effective models* for the learners whose development they are expected to further.

In grasping the significance of these top two phases of the Framework, parents and teachers of children growing right now will glimpse the ideal outcome of their nurturing and teaching at all of the preceding stages of development. That nurturing and teaching, if conducted within optimal circumstances, will have served to advance the young ones benefiting from it systematically from Phases One through Five.

In grasping the significance of these top two phases of the Framework, parents and teachers of children growing right now will glimpse the ideal outcome of their nurturing and teaching at all of the preceding stages of development.

To reiterate, the intelligence emerging at each stage of optimal human development expresses a corresponding aspect of what we have termed Nature's Plan, and can be facilitated by Guiding Strategies geared to that particular developmental stage.

Keep in mind also that sufficient mastery of the developmental tasks at each stage is required to ensure that future stages can be completed equally successfully. With that in mind, I will discuss aspects of Emergent Intelligence at Phases Six and Seven, relate these aspects to the corresponding stages of Nature's Plan, and specify Guiding Strategies that can facilitate progress through these higher stages.

Regarding the lower half of the Framework table, we see that Emergent Intelligence at Phase Six (labeled "Global-Systemic") entails intuition, insight, integration skills, and meaning-making at the highest levels. These emergent abilities reflect Nature's Plan at this level, which is to connect or identify oneself with a comprehensive ontological reality that transcends a limiting sense of self.

As for Guiding Strategies during Phase Six, an immense variety of such strategies is available, some of them having already been discussed in *Anatomy of Embodied Education*. They include a range of methods such as psychosynthesis for training mindfulness, plus biofeedback and neurofeedback technologies to assist in mind/body integration, plus many other methods for expanding and refining consciousness that have been around for a long time and are well-tested.

As implied earlier, I assume that many readers of this book have already reached this developmental stage, and may even be functioning as teachers and guides who assist Phase Five learners in entering and navigating Phase Six. Other readers who are just reaching this stage might want to consider the current discussion a guide to what is possible for them after having moved beyond their current stage of development.

The Pinnacle of the Model; Phase Seven

With Phase Seven I approach the culmination of the developmental model encompassed by this book. At this stage, descriptors depart from the realm of neuroscience and become more spiritual-sounding. Transpersonal and Transrational, for instance, are terms suggesting emergent intelligence beyond the scope of what is ordinarily cultivated in schools. Phase Seven is included in this presentation for the sake of completing the developmental portrait being painted here, and to suggest what is possible for human learners whose innate capabilities for balance, integration and adaptation are carefully recognized and brought forth by institutions responsible for educating them.

Of the pioneer developmental theorists surveyed in the Framework, one in particular stands out as a paragon for me. I speak of Abraham Maslow, whose socio-phenomenological research in the mid-20th century yielded what for many of my generation was our first serious glimpse of the possible human, evolved beyond "adjustment" to a regressive culture. Maslow is widely recognized as a founding father of humanistic psychology, which in turn led to transpersonal psychology, and hence to the more recent derivation, positive psychology. His work reached its pinnacle in the 1960s, and was taken up by a host of successors following his death in 1970.

I have chosen to conclude the main text of this book with a brief overview of Maslow's seminal discoveries and ideas, because they encapsulate my motivation for writing it. That motivation revolves around the questions spurred by the two peak experiences I described in the introduction/overture. Also pertinent to my aims are some of his remarks about the relationship of culture to the highest reaches of human development. These will be discussed below.

Maslow is best known for contributing three seminal concepts to the foundation of studies in human potential:

1. *the hierarchy of human needs* (partially summarized following his name in the Framework table);

2. *self-actualization* (at the apex of the hierarchy);

3. the phenomenon of *peak experience* (which he considered the climactic aspect of self-actualization).

His final book, *The Farther Reaches of Human Nature* (1971), is an edited collection of essays detailing the development of these three concepts and others closely related to them.

Maslow's hierarchy, number one in the above list of his developmental contributions, figures prominently in the model presented here. As for the third concept listed above, peak experience, its meaning and implications are suggested by the Phase Seven descriptors in all three rows in the lower half of the Framework.

This entire model is predicated on a progression of human development such as that studied by Maslow and the others shown in the table. And just as self-actualization is uppermost in Maslow's hierarchy of developmental needs, its equivalent in this model represents the pinnacle of human development as it is currently understood.

The current model not only incorporates the work of five pioneering theorists, it serves to extend and validate their developmental theories by adding knowledge about neural development that, as I emphasized earlier, was unavailable to them during the time they were forming those theories.

Yet even without that knowledge, one of them (Maslow, in his study of self-actualizing subjects) had discovered a characteristic common to his subjects that he considered a hallmark of psychological health. In self-actualizing people, he found, "...dichotomies were resolved, polarities disappeared, and many oppositions thought to be intrinsic merged and coalesced with each other to form unities" (Maslow, 1954, p. 233). As examples of such dichotomies he listed

> "...kindness-ruthlessness, concreteness-abstractness, acceptance-rebellion, self-society, adjustment-maladjustment... serious-humorous... introverted-extroverted, intense-casual... mystic-realistic, active-passive... a thousand serious philosophical dilemmas are discovered to have more than two horns, or, paradoxically, no horns at all." (pp. 233-234)

Now fast-forward several decades. J. A. Scott Kelso (2008) was quoted in Chapter 4, saying "...coordination dynamics opens up a path to reconciling contradictions, dualisms, binary oppositions, and the like in all walks of life, illuminating thereby their complementary nature..." (p. 185).

The parallel between findings by Maslow, quoted two paragraphs back, about characteristics of self-actualized people he had studied, and Kelso's conclusions a half-century later about the dynamics and effects of coordination dynamics, is quite striking. For me the congruence between results of two research methods so utterly distinct from each other strongly affirms the validity of the conclusions drawn from both sets of results.

To clarify further, my understanding of that particular congruence is that the pinnacle of healthy psychological development reported by Maslow—self-actualization—includes transcendence of duality, and the pinnacle of neurobiological functioning worked out by Kelso—balanced coordination dynamics—also features transcendence of duality.

From my early twenties on I have felt strongly that one of the most troublesome instances of duality in western culture is the duality of mind and body, famously set in stone by Descartes hundreds of years ago. That false dichotomy has led Western thinking astray for all those centuries. Surely the recognition of the value of transcending duality in general— of transmuting it instead to something like complementarity—surely that recognition is a milestone in our understanding of the highest level of healthy human development. Surely the demonstration above that this recognition spanned at least two generations of careful observation and thought, from Abraham Maslow to J.A. Scott Kelso, contributes to its validation.

With regard to the mind-body duality in particular, its transcendence has been furthered also by the discussion in previous chapters of, for example, Daniel Siegel's work in interpersonal neurobiology. In that work, mind, body/brain, and relationship/environment are explicitly intertwined. In fact, that transcendence is an implicit undercurrent throughout this book.

Returning to Maslow and *The Farther Reaches of Human Nature*, the following excerpts must be carefully considered:

> Self-actualization is not only an end state but also the process of actualizing one's potentialities at anytime, in any amount. (p. 47)

> Peak experiences are transient moments of self-actualization. They are moments of ecstasy which cannot be bought, cannot be guaranteed, cannot even be sought... but one can set up the conditions so that peak experiences are more

likely, or one can perversely set up the conditions so that they are less likely. (p. 48)

The two statements quoted above are admirably lucid, and move my understanding of both self-actualization and peak experiences significantly forward. Then, much further along in the book, in the chapter titled "Goals and Implications of Humanistic Education," he states:

We know that children are capable of peak experiences and that they happen frequently during childhood. We also know that the present school system is an extremely effective instrument for crushing peak experiences and forbidding their possibility. (p. 188)

The above assertion that "children are capable of peak experiences and that they happen frequently during childhood," coupled with the two quotes above that, demonstrate that Maslow's thinking about self-actualization and peak experiences developed considerably from the time he began his exploration of self-actualization in the mid-20th century, when he was studying only exemplary adults, to his understanding of it toward the end of his life.

Thus, as I interpret that development, *self-actualization is a process that is possible during any of the seven phases of human development described in Chapter 7!* We don't have to wait for mature adulthood to experience it. That means that a person of any age can, given sufficient attention to Nature's Plan as described in the Framework, experience fulfillment that is potential at that particular age. Therefore a human of any age can self-actualize, and age-appropriate peak experiences can emerge, during those phases. They are not relegated solely to well-developed adults, although certainly the nature of those experiences will change from one phase to the next all the way to the end of one's life. I use the above term *emerge* deliberately, invoking its meaning within complex dynamical systems theory.

Understanding it this way enables me to comprehend why I was able, as a young adult, to undergo the blissful peak experiences I described in the introduction to this book. I return now to the questions that have emerged from those experiences, and how I can respond to them in light of what has been presented in this book so far.

Some questions from the initial experience:

- *What in the nature of humans gives us the capacity to awaken into such joy?*

Daniel Siegel (2017) wrote: "a turning point in one's life may arise when the top-down filters that shape our feelings, perceptions, thoughts, and actions are suddenly broken down and shaken up, and a new bottom-up experience fills our awareness" (p. 142). I feel certain that the "bottom up experience" to which Siegel referred can include joy at the level I felt during my initial peak experience.

Bolstering my certainty are Siegel's remarks about an experience he had as a college undergraduate that corresponds very closely to the one I have described. The precursor of his experience was radically different from mine: he had suffered injuries, including a head trauma that resulted in a day-long loss of personal identity. "I was wide-awake," he wrote, "but had no idea who I was." Also corresponding to my experience, Siegel "…was filled, moment-by-moment, with sensory immersion that felt, well, somehow complete. There was nothing missing, nowhere to go, and nothing else to do but let the experience flow" (p. 126).

He concluded from this experience and its aftermath: "You can be awake—perhaps even more fully awake—if your personal identity with all its baggage of history, learning, judgements, and filtering of perception is suspended." Much of what he learned concerning "top-down and bottom-up" information processing stemmed from this experience. There is value, he suggests, in tempering and even occasionally suspending the filtering effect of the top-down processes

Exploring how those top-down filters get suspended is a very worthwhile project. That exploration might bridge to ancient wisdom traditions such as Sufism and Zen Buddhism, and also intersect with attentional skills such as Open Focus, and with other mindfulness practices currently being adopted by many schools.

The information presented below with regard to the next question, and my reflections on the information, I offer as a step toward the exploration.

- *What had I needed to transcend in order to experience that blissful absence of identity?*

Neuroscience and Michael Pollan's book on entheogens (2018) bring the default mode network (DMN) to the fore in answer to this question. The DMN in human brains was discovered by brain researchers relatively recently—near the beginning of this century—using state-of-the-art imaging technology to measure aspects of neural activity. It is a network of distinct neural structures in various regions of the cerebral cortex. These regions (which bear exotic names such as medial prefrontal cortex, ventral precuneus, posterior cingulate cortex, and parts of the parietal cortex) oscillate synchronously together under certain conditions, certain states of consciousness. Under other conditions of consciousness, the DMN goes quiet. The distinction between conditions that activate the network and conditions that de-activate it is of paramount interest in this query. It pertains directly, as I shall now elucidate, to the question posed in italics above.

Research has repeatedly shown that when one's brain is actively engaged with the external world, taking in and processing sensory information, only certain parts of it, certain neural circuits, are active and using energy provided by oxygen and nutrients in the bloodstream. But when the external world ceases to claim attention, those circuits become inactive. That is when the DMN—the default mode network that is altogether different from the externally focused circuits— goes to work focusing on internal processes such as thinking about the past, worrying about the future, sorting out thoughts, and creating mental constructs.

The mental constructs created by the DMN generally act as "top-down filters" that govern our perceptions of what exists in the environment both outside and inside our skin. (These filters were introduced in the Siegel quote on the previous page.) From an evolutionary standpoint they are considered mostly beneficial and adaptive, in that they a) incorporate life experience to form remembered patterns that b) influence how we respond to familiar situations, with c) the goal of maximizing functional engagement while d) making the most efficient use of biological energy.

In other words, the DMN is developed, beginning in childhood, by learning from the experience of living.

Classroom learning is a subset of overall learning. When guided skillfully and humanely, classroom learning can powerfully augment and guide learning outside the classroom. Conversely, as discussed extensively in Chapter 3, the classroom experience can have negligible or even negative effects on extra-curricular life experience.

This can be the case especially when it serves only to promote regressive, corrosive societal values and practices. Thus, although it is true that DMN-generated constructs can be beneficial and adaptive, it is also true that those constructs can constrain and distort one's lived experience.

Apparently, the most important mental construct arising within the DMN is the one generally known as the ego or self—the everyday, ubiquitous sense of personal identity. It is now time to explore the vital, intimate linkage between the neurological default mode network and the psychological self.

Pollan (2018), who uses the term "top-down" in the same context as does Siegel, wrote: "As a whole, the default mode network exerts a top-down influence on other parts of the brain, many of which communicate with one another through its centrally located hub" (p. 303). He had gleaned information about the DMN by interviewing scientists such as Robin Carhart-Harris, who had conducted government-sanctioned research in the UK studying the effects of psilocybin on the brain and mind.

Carhart-Harris, using functional magnetic resonance imaging (fMRI) technology, had tracked neural network (including the DMN) activation and de-activation in participating volunteers' brains under the influence of the psychedelic substance psilocybin. The volunteers' verbal reports of their experiences while under the influence of psilocybin were recorded at the same time that the brain scans were underway.

The comparison between the scans and the reports revealed that

...the steepest drops in default mode network activity correlated with his volunteers' experiences of "ego dissolution." [emphasis added]...The more precipitous the drop-off in blood flow and oxygen consumption in the default mode network, the more likely a volunteer was to report the loss of a sense of self. (p. 304)

For me, this finding conclusively demonstrates the neurophysiology underlying my spontaneous experience of ego dissolution, the first peak experience described in the Introduction. It is the explanation I have been seeking ever since. And the same explanation applies to another study conducted soon after Carhart-Harris had published the results of the research described above.

The more recent study, done at Yale University by Judson Brewer, also used fMRI to assess neural network activation. But instead of studying the effects of a psychedelic substance, Brewer was peering into the brains and minds of experienced meditators. Pollan, reporting on Brewer's study, writes that the Yale scientist

> ...noticed that his scans and Robin's [i.e., Carhart-Harris's] looked remarkably alike. The transcendence of self reported by expert meditators showed up on fMRIs as *a quieting of the default mode network* [emphasis added]. It appears that when activity in the default mode network falls off precipitously, the ego temporarily vanishes, and the usual boundaries we experience between self and world, subject and object, all melt away. (p. 305)

On the same page, Pollan continues:

> This sense of merging into some larger totality is of course one of the hallmarks of the mystical experience; our sense of individuality and separateness hinges on a bounded self and a clear demarcation between subject and object But all that may be a mental construction, a kind of illusion...

That last passage above precisely characterizes both of the peak experiences I recounted in the Introduction. The first one certainly entailed "ego dissolution," whereas the more recent one, enabled by ingesting LSD, featured more of "the hallmarks of the mystical experience" referred to by Pollan. Thus my two experiences bracket the range of possibilities demonstrated by these two very important studies—the possibilities inherent in temporarily suspending activation of the default mode network *while simultaneously foregoing the activation of narrow-focused attention toward the world "outside."*

To clarify further: in the initial experience I awoke from deep sleep in a relatively unfamiliar environment with my sense of identity (i.e. self, ego) completely absent, with my perception of my surroundings vivid and unconditioned by a sense of their being "my" perceptions. They existed within an attentional field that would be characterized by Les Fehmi as purely "open focused." Additionally, there was nothing in that situation that prompted the delighted being "I" had become to *do* anything other than enjoy the perceiving; I felt no need to *go* anywhere else, to think or remember or plan—none of the myriad actions that would have quieted my brain's DMN, but would also have blocked the extraordinary experience unfolding at that moment.

The second experience also entailed ego dissolution, with the added dimension of immersion in a vastly expanded field of existence. In that field, time did not exist, nor did any whisper of a need to engage with a three-dimensional world.

In both of those highly unusual circumstances, perception of both "outer" and "inner" environments emerged purely, unconstrained by conditioning. The first circumstance was simpler and briefer; the second was…something else.

We come now to the final question engendered by the initial experience.

- *What conditions enabled me to experience such bliss on that particular occasion?*

I was in completely new territory emotionally: joyfully immersed in love, and simultaneously stricken by regret as I regarded the sudden and prolonged absence of the young woman who had inspired this love. This very dynamic juxtaposition of contrary emotions confronted me with a dualistic paradox that I could resolve only by shifting into a non-ordinary state of consciousness (harken back to Maslow's characterization of self-actualizers as transcending dualities). Reflecting on it now, I realize that the ego-free state of consciousness that presented itself was permeated by *transpersonal* love. On the other hand, if I had remained saturated with a sense of personal selfhood, my state would likely have been beset by regret and sadness.

In CDST terms I had reached a bifurcation point: *love and bliss* constituted one attractor, *regret and sadness* another. Recognizing that the unconscious (or is it superconscious) choice tipped toward love rather than sadness teaches me something hopeful about myself, and potentially about humanity in general. As naïve as I was at that stage of my life about transpersonal phenomena, I still awoke into a peak experience. If that could happen to me, it could likely happen to most anyone—and perhaps it does, more often than is commonly known. Given a culture such as ours in which such experiences are either unknown or undervalued, it is unlikely that people would speak openly about them. The issue of culture is addressed in the next section.

But first, some final remarks about the second set of questions, that emerged from the LSD journey:

- *How can human consciousness encompass the magnitude of existence that I experienced at the outset of that journey?*

- *Is it possible to develop, without chemical entheogens, the capability to experience the cosmos to that extent, and to apply that capability to living day-to-day?*

- *How might humans benefit from applying such capability to everyday existence?*

Clearly these are far-reaching questions, worthy of a depth of exploration that is beyond the scope of this book to undertake. However, a beginning of that exploration may be glimpsed by revisiting Abraham Maslow's emphasis on peak experiences. As we have seen in this chapter, Maslow considered peak experiences an integral aspect of the developmental phase he named self-actualization

But, as discussed earlier, he did not consider peak experiences to be limited *only* to the most advanced phase of human development. We see this, for instance in Chapter 12 of his book *The Farther Reaches of Human Nature,* which is titled "Education and Peak Experiences." In that paper, Maslow made it clear that he considered it an obligation for schools to provide settings for, and to encourage students to engage in, *age-appropriate* peak experiences.

I hope you will keep his mandate firmly in mind as you reflect on the developmental model presented at the beginning of the current chapter. It means that not only are peak experiences a hallmark of the highest phase of human development (Phase 7 of the model, self-actualization), but that the equivalent of this hallmark can be experienced at any of the preceding developmental phases, provided that each phase of a life's journey is effectively navigated.

The relative quality of peak experiences will of course increase with each succeeding phase of development. Maslow urges us as educators to recognize the nature of that development—know what is possible and optimally beneficial—and facilitate its progress. Knowing and facilitating what is possible and optimal is basic to bringing forth what I have termed *true intelligence.*

And when I think of the potential outcome of knowing and practicing the pathways to true intelligence, I get a glimpse of answers to the the deep questions posed on the previous page. Those questions emerged, after all, from a transcendent peak experience in my own life.

Addressing Our Culture

In the preceding seven pages I have elaborated upon my current understanding of Abraham Maslow's pioneering explorations in human development beyond mere adjustment to a world too narrowly conceived. I've described how my understanding has been enriched by reflecting on peak experiences in my own life, and couching those experiences in terms of contemporary neuroscience. In writing these pages my purpose has been to augment my advocacy for root changes in the widely-held purposes and practices of education now extant in the US and elsewhere—changes that would enable peak experiences (and the self-actualizing process from which they emerge) to be more commonly accessible.

What I wish to do now, in closing out this final chapter, is share with my readers some of Maslow's understanding about the *cultural context of self-actualization*. You will see as we go along the relevance of his insights to the challenging project of applying the material in this book to the evolution of education and parenting.

Maslow (1971) equated self-actualization with what he called "the value-life (spiritual, religious, philosophical…)," and insisted that such a life "is an aspect of human biology and is on the same continuum with the 'lower' animal life (rather than being in separated, dichotomized, or mutually exclusive realms)" (p. 324). Because such a life is a biologically-based human potential, *any person on the planet might grow to live a value-life, might become a self-actualizing human being.*

Let me remind you that the inclusion of brain maturation stages alongside the developmental stages formulated by Maslow (and the others) supports what for them could only be a conviction based on their analyses of behavioral observations, lacking the abundance of correlative neuroscientific knowledge presently available to us. He had intuited what has now been solidly supported by neuroscience—that the value-life exists on a continuum of neurobiological development from the very earliest stages to the highest, and therefore is potentially attainable by anyone.

He warned however that such an achievement "must be actualized by culture in order to exist" (p. 326).

Human motivations at the level of self-actualization, at the level of living a value-life, involve the satisfaction of what he termed *metaneeds*, and he therefore collectively termed them *metamotivation*. He recognized that, although the potential exists for every human to be motivated at this level, they "can become widely actualized *only* in a culture which approves of human nature, and therefore actively fosters its fullest growth" (p. 327).

> Culture is definitely and absolutely needed for their actualization; but also culture can fail to actualize them, and indeed this is just what most known cultures actually seem to do and to have done throughout history. (p. 326)

A culture can be judged, he said, "in terms of the degree to which it fosters or suppresses self-actualization, full humanness, and metamotivation" (p. 326).

Reading this blunt statement is bound to engender some degree of discomfort in many of us. It challenges us to regard our own culture in light of whether "it fosters or suppresses self actualization, full humanness…" In cultures currently existing in the Western Hemisphere, many indicators may be found to support a pessimistic assessment in response to this challenge. To the extent that one's culture suppresses self-actualization and full humanness, it would obviously be challenging to create an educational setting consistent with the purposes of this book.

However, there is reason for hope. To counter any pessimism aroused by the challenge, let us recognize that no culture is monolithic—that is, all one way. Recall the statements in the previous chapter warning against the tendency to reify phenomena, to think of them as "things" that can be objectified, classified, and thereby kept at an intellectual distance. I urged instead (along with Bateson, Siegel and others) that phenomena be experienced and considered as dynamic *processes* rather than "things." It is the same for cultures: they can best be regarded as dynamic processes, as systems in flux.

Much is now known about how to influence processes like this. Close examination of their dynamics will often reveal "tipping points," nodes in their unfolding where change may be induced without great effort. I propose that the culture of education in the West may be regarded in this way. Any deficiencies that currently exist in its ability to foster balance, wholeness, self-actualization, full humanness, are not necessarily set in stone. A thorough familiarity with the systems concepts presented throughout this book can empower readers to regard apparent cultural impediments to full humanness with new eyes, and thereby discover means for moving beyond those impediments. Understanding concepts such as self-regulation, self-organization, complexity, emergent properties, attractors, bifurcation and the like can reveal means to overcome those deficiencies.

To state my conclusion bluntly and succinctly: if the developmental phases described in this chapter are understood clearly enough, and valued highly enough, *ways can be found or invented to bring them to fruition.* Keeping in mind the repeated emphasis in this book on relationship as one of the primes of the learning human mind, you are encouraged to seek and create such ways—not in isolation, but in conjunction with other seekers and creators.

Explore and ponder the model. Find ways to apply it. Share it with friends and colleagues. Enlist their help in understanding, applying, and even expanding it. *Engage in first-hand experience with mindfulness practices such as Open Focus training,* for only by this personal engagement will you fully grasp the necessity for incorporating the essence of these practices—along with the knowledge base presented in this book—in the project of evolving education.

Let's work together to grow a Phase Seven culture: a culture in which the new normal is self-actualization.

References

Benard, B. (1991). *Fostering Resiliency in Kids: Protective Factors in the Family, School and Community*. Western Regional Center for Drug-free Schools and Communities.

Erikson, E. (1963). *Childhood and Society*. Oxford University Press.

Goleman, D. (2006). *Social Intelligence: The New Science of Human Relationships*. Bantam.

Kelso, J. A. S. (2008). An Essay on Understanding the Mind. *Ecological Psychology*. 20(2), 180-208.

Kohlberg, L. (1984). T*he Psychology of Moral Development: The Nature and Validity of Moral Stages (Essays on Moral Development, Volume 2)*. Harper & Row.

Maslow, A. (1971). *The Farther Reaches of Human Nature*. Viking Press.

Piaget, J. (1952). *The Origins of intelligence in Children*. International Universities Press.

Pollan, M. (2018). *How to Change Your Mind*. Penguin Press.

Siegel, D. (2017). *Mind: A Journey to the Heart of Being Human*. W. W. Norton & Co.

Steiner. R. (1996). *The Education of the Child in the Light of Spiritual Science*. Anthroposophic Press.

Coda

Systems Perspectives on COVID-19 and Beyond

Just as I was approaching completion of this book, and beginning to explore possible avenues to its publication, a historic event occurred: the outbreak of the COVID-19 pandemic. As the pandemic progressed, provoking desperate measures by authorities around the world to contain it, it began to dawn on me that I could not consider the book complete without observing responses to this catastrophic event and its ongoing consequences, and framing my observations in the context of the systems models presented in the book.

With this realization, I began to search the Internet for any writing that linked worldwide responses to the pandemic on one hand and complex dynamical systems theory on the other. It did not take long for such material to begin showing up, and within a period of a few months it seemed to me that widespread speculative and analytic writing on this subject was quickly taking on the stature of a new journalistic genre.

While I was collecting and assimilating relevant publications, another catastrophic event occurred—one tragically not as historic as governmental attempts to contain the pandemic, in that similar events have been happening for a long time. I refer to the on-camera killing of a black man by a white policeman while the victim was restrained by handcuffs and lying face-down on a street in Minneapolis, Minnesota USA.

As I suggested above, such killings already comprise a sickening strand in the tapestry of racial oppression in the country in which I was born. These killings often provoke outrage, demonstrations, and occasional ineffective attempts to reform police practices.

But the murder of George Floyd, viewed on video screens around the world, had an effect surpassing any that I had ever before witnessed: the prompt, orderly mobilization of vast numbers of protest demonstrators in the US and beyond. Unexpectedly, following this eruption of protest, an incident was recorded of a county sheriff walking with and listening to a group of demonstrators, after having laid aside his weapons and riot gear. I had never witnessed such an act of compassion in that setting, and was briefly flooded with hope that it would become a liberating meme.

Instead, the throngs of demonstrators were frequently infiltrated by provocateurs, who successfully drew often brutal shows of force from police apparently unable or unwilling to distinguish peaceful demonstrators from opportunistic looters and tactically sophisticated gangs. In July 29, 2020, the newspaper *USA Today* reported:

> Two days after the death of George Floyd, a surveillance video showing a man dressed in all-black garb and holding a black umbrella breaking the windows of an AutoZone with a sledgehammer went viral.

> The individual…was a member of a white supremacist group with intentions to stoke racial tension, police confirmed Tuesday.

> Before breaking the windows, he spray painted "free (expletive) for everyone zone" on the front doors of the shop. Some protesters confronted the man to stop.

> Hours later, a fire broke out at the store — the first that firefighters responded to during the protests that took place May 27.

> The revelation, first reported by the *Star Tribune* in Minneapolis, confirms the suspicions of some protesters that unaffiliated agitators sought to cause discord in the protests.

Donald Trump responded by compounding the chaos, attempting to deploy US military forces against US citizens exercising their constitutional rights.

He did not immediately succeed in this because military leaders bluntly refused to comply with illegal orders from their immediate "superior officer" the Commander-in-Chief. (If there is a historical precedent to this outright refusal of the highest-ranking military establishment to comply with or condone an illegal order from the President of the United States, I do not know about it.)

But he has persisted in his attempts: as I write this revision, many news articles have described militarized federal agents patrolling a major city in unmarked vehicles, assaulting demonstrators and taking some into custody with apparent disregard for established legal process. Trump has threatened to repeat this caustic, dangerous strategy in other major cities, and as of the time of this writing, has begun following through.

My purpose in describing these momentous occurrences is to illustrate chaos, on a thunderingly large scale, rampaging across a broad swatch of socio-cultural phenomena. Repeatedly throughout this book I have asserted that systems theory is an exceptionally effective tool for modeling such phenomena, so that they may be understood more deeply and potentially addressed more effectively as a result of that understanding. It is difficult to imagine a more stringent test of my assertion than to apply systems theory—especially complex dynamical systems theory (CDST)—to aspects of the current world crisis, and find out a) to what extent the theory deepens understanding of those aspects, and b) whether it might help to guide not just recovery from these crises, but actual evolutionary development resulting from them.

Accordingly, I have gathered, from both print and Internet sources, a sample of written commentaries concerning COVID-19 and subsequent crises, with the intention of summarizing this material and then framing it in terms of systems theory. Not only will this exercise serve as an example for readers of how systems theory can be pragmatically applied; it might also contribute to the general understanding of how humankind can not only get through these crises, but actually grow from them.

Material Sampled and Summarized

"The Post-Corona World" by Matthias Horx

Horx begins this article by stating his major premise, which entails a cardinal feature of complex systems:

> *At the moment I am often asked when Corona 'will be over' and when everything will return to normal. My answer is: never. There are historical moments when the future changes direction. We call them bifurcations. Or deep crises. These times are now.*

(Note: "bifurcation" is explicated in Chapter Three, e.g.: "In the parable, our pilgrim's state space reached a level of instability at which a competing attractor might force the trajectory to bifurcate, to swerve toward this new attractor.") Horx suggests:

> *The world as we know it is dissolving*—[we can consider this an extreme "level of instability!"] *but behind it comes a new world* [i.e. "a competing attractor"], *the formation of which we can at least imagine* [thus enabling us to "swerve toward this new attractor"].

He then offers an example of how we can imagine what this new world might manifest...

> *Let's imagine a situation in autumn, let's say in September 2020. We are sitting in a street cafe in a big city. It is warm and people are walking down the pavements again.*

...and then imagine viewing the new world from a future vantage point:

> *...Looking back, what will we be surprised about? We will be surprised that our social distancing rarely led to a feeling of isolation. On the contrary, after an initial paralyzing shock, many of us were relieved that the constant racing, talking, communicating on a multitude of channels suddenly came to a halt. Distancing does not necessarily mean loss, but can open up new possibilities. The social courtesy that we previously increasingly missed, increased.*

The range of surprises widens:

> ...Teleconferencing and video conferencing...turned out to be quite practical and productive. Teachers learned a lot about internet teaching. The home office became a matter of course for many — including the improvisation and time juggling that goes with it...Reading books suddenly became a cult...Reality shows suddenly seemed awkward and the whole trivia trash, the garbage for the soul that flowed through all channels seemed ridiculous. The exaggeration and culture of fear and hysteria in the media was limited after a short first outbreak.

He proposes a number of things we will have witnessed in this imagined future:

> A crucial change in social behaviour...people could have solidarity and be constructive despite radical restrictions. Human-social intelligence has helped. The much-vaunted artificial intelligence, which promised to solve everything, has only had a limited effect on Corona.

> This has shifted the relationship between technology and culture. Before the crisis, technology seemed to be the panacea, the bearer of all utopias. No one — or only a few hard-boiled people — still believe in the great digital redemption today. The big technology hype is over. We are again turning our attention to the humane questions: What is mankind? What do we mean to each other?

After discussing a range of other possible outcomes, Horx returns to the "humane questions," and imagines:

> In the new world, wealth suddenly no longer plays the decisive role. Good neighbors and a blossoming vegetable garden are more important. Could it be that the virus has changed our lives in a direction that we wanted to change anyway?

> In the middle of civilization's shutdown, we run through forests or parks, or across almost empty spaces. This is not an apocalypse, but a new beginning.

He concludes with this somber overview:

Human civilization has become too dense, too fast, and overheated. It is racing too fast in a direction in which there is no future.

In phrasing introduced in Chapter 3, Horx's overview means that the system ("human civilization") is far from equilibrium, rendering it susceptible to bifurcation. He goes on:

But it can reinvent itself. System reset. Cool down! Music on the balconies! This is how the future works.

In dynamical systems terms, Horx has imagined and characterized the new "attractor."

* * *

"Can the Lockdown Push Schools in a Positive Direction?"

Here Are Five Ways That COVID-19 Could Change Education for the Better." by Patrick Cook-Deegan (associated with The Greater Good Science Center at the University of California, Berkeley), May 20, 2020.

I was particularly gratified to find Cook-Deegan's commentary because of its direct relevance to the theme of education that is central to this book. The author begins by immediately and dramatically setting the stage:

The COVID-19 crisis has closed over 124,000 schools in America. Most will be closed until next fall, with many likely experiencing roving blackouts throughout the year. Since the rise of compulsory schooling in America a century ago, there has never been this level of school shutdown. Not during the Spanish Flu of 1918 or World War II, or after 9/11.

He succinctly summarizes procedural changes that were adopted early on:

Teachers and schools are doing their best to adjust to this strange new land. Zoom classes, asynchronous learning, and Facebook lessons are all being implemented with varying degrees of success.

Then comes the big question about potentially "fundamental" changes in the paradigm of education:

> But what happens when school comes back in session? Are these closures merely a "long snow day," as one educator put it? Or will this experience fundamentally change the nature of what it means to "do school"?

> Leading political scientists and sociologists have documented that most large-scale social changes happen amid and in the wake of crises…COVID-19 certainly counts as a large-scale crisis, hence the opportunity for transformative social change to education.

Cook-Deegan discusses several arenas within which the the turbulence resulting from COVID-19 could engender fundamental and much-needed changes in American education:

> There are five ways I believe that COVID could change the future of school—for the better.

> 1. _More social-emotional learning (SEL) for students_…There's widespread acknowledgement that we must pay greater attention to the social-emotional needs of our students because they're suffering. When we get back to school, teachers and students will have to process their parents' lost jobs, their tough times with their families at home, and how this crisis affects their future when it comes to college. If school resumes and this work isn't prioritized, students will feel like schools really don't get it and are out of touch with their needs.

> 2. _Higher priority on teacher well-being_…Of course, this crisis is hitting teachers hard, too. A Yale Center for Emotional Intelligence survey of 5,000 teachers amid COVID asked them to describe the three most frequent emotions they felt each day…Anxiety, by far, was the most frequently mentioned emotion, according to the study… The primary sources of teacher frustration and stress were feeling unsupported by their administration regarding challenges related to meeting students' diverse learning needs, high-stakes testing, an ever-changing curriculum, and work/life balance.

3. *More of a coaching and mentoring role for teachers...I've argued for years that educators need to be trained more like mentors and coaches and less like knowledge dispensers and disciplinarians...mentors share more of themselves and who they are, and understand their role as providing support and encouragement, rather than just keeping students disciplined or moving along in the curriculum...Over the years, creative school models, like Big Picture Learning, have popped up that intentionally foster mentoring relationships. Students work closely over multiple years with one "advisor" who works with each student and their family to craft individual learning plans. Students also engage in internships at community organizations and businesses where they develop relationships with mentors as they explore career possibilities. Multiple studies have suggested that this approach leads to increased student engagement and stronger interpersonal and intrapersonal skills such as collaboration, self-efficacy, and academic engagement.* (Note the levels of development, as portrayed in the Chapter 7 table "Framework for Optimal Human Learning and Development," that are spanned by such skills and experiences. Innovative models such as Big Picture Learning offer interesting alternatives to the standard US model critiqued in Chapter 1.)

4. *More autonomy for schools and teachers, fewer top-down demands... In light of pandemic stay-at-home orders, the fed and many states have dropped year-end testing requirements. Many of these tests were implemented as well-intentioned albeit poorly designed public policy measures to force accountability for schools and close the equity gap for students of color.*

   However, more than 15 years since the inception of No Child Left Behind, there's almost no evidence that these tests accomplish either of these two objectives. Nor is there evidence that these types of tests actually help students learn the material. Yet schools spend a significant amount of their school year prepping for these tests, most school leaders loathe them, and teachers cite them as a major reason for leaving the profession. So, what if we seize this moment to scrap most of these tests and find other ways of measuring proficiency, and free up school time for actual learning?

5. _More student choice and autonomy_...One well-cited study[9] showed that students take in knowledge best when they're intrinsically motivated to learn something. Because of COVID, demand has increased for programs like Outschool, where students opt in to an online teaching class with a teacher in small class settings...(in which the teachers must want to teach the content and the students must opt-in to the program. In other words, there's more intrinsic motivation on both sides for these classes, which leads to better results...In the COVID era..."teachers have the opportunity to explore online teaching platforms, rethink the purpose of learning, and design activities that engage students in a whole new way," said Katie Barr, the principal of Maria Carillo High School in Santa Rosa, CA. "It can often be challenging to make change amid normal political circumstances...COVID is a chance for us to fundamentally rethink our system." Barr said. "COVID is presenting a unique opportunity in education. For the first time in 150 years, we get to blow up the industrial model of education. We are given the gift of learning because we want to learn..."

This commentator concludes:

> Once stay-at-home orders are lifted, students at more traditional schools might chafe at coming back to the less autonomous model of schools and fight for more academic freedom. In addition, teachers may not want to go back to the set curricula they had to follow before. Crisis breeds disruption and innovation—and often creates a future that was _possible_ before but _impractical_ pre-crisis. In other words, once people experience something different, it can often be hard to put the genie back in the bottle...If there was ever a catalyst to jumpstart change we are living in it right now.

* * *

9   Cordova, D. I. and Lepper, M. R., Intrinsic Motivation and the Process of Learning: Beneficial Effects of Contextualization, Personalization, and Choice; _Journal of Educational Psychology_, 1996, Vol. 88, No. 4.

"The Light at the End" by Nafeez Ahmed; May 11, 2020; *YES! MAGAZINE*, pp. 19-22.

Dr. Ahmed, who directs the System Shift Lab and is Research Fellow at the Schumacher Institute for Sustainable Systems, has applied his thorough knowledge of systems thinking to the writing of this impressive article. After a descriptive preamble, he launches into teaching mode by placing viruses generally in an evolutionary context, then making a surprising statement about COVID-19 in particular:

> ...*Viruses have an environmental function as an evolutionary force for biological organisms. Recognizing this allows us to reframe our understanding of the pandemic—which neither comes out of the blue, nor can be simply defeated using the instruments of advanced medical science. <u>On the contrary, the pandemic has been incubated by the very structure of our civilization.</u>*

To justify this assertion, Ahmed refers to a "paper in the journal Proceedings of the Royal Society B," summing it up as follows:

> ...*the very same process of industrial expansion behind the collapse of biodiversity (the variety and abundance of life on earth)—putting a million species at risk of extinction—is also responsible for the heightened risk of major disease outbreaks.*
>
> *The global expansion of human activities, the study finds, has caused escalating "losses in wildlife habitat quality," leading to "increased opportunities for animal-human interactions." These in turn have "facilitated zoonotic disease transmission"—that is, the jumping of diseases found in certain animals to human populations.*

He widens his analysis of the pandemic's effects to encompass the likely collapse of what he terms "the global system," which he characterizes as "structurally hardwired for endless exponential economic growth simply to retain stability." His bleak conclusion:

> *The pandemic, triggered by the operation of the global system, has placed that system between a rock and a hard place, where every option signals the inevitability of a long-term economic*

contraction of some kind. The human species has hit a roadblock—a structural impasse of our own creation.

As a widely-published journalist he has reported extensively on the growing economic disruption within a variety of industries—primarily extractive, energy-related industries—and summarizes enough of that information in this article to support these incisive statements:

> *When understood in its full global systemic context, the COVID-19 crisis reveals the fundamental limits of the paradigm that defines industrial civilization in its current form—its assumptions about human nature, conception of the natural world, economic theory of the relationship between the two, overriding value-system, and associated nexus of collective behavioral patterns.*
>
> *While this paradigm has brought us this far, it has thrown us into an unprecedented crisis exposing the levels of cognitive dysfunction baked into our current system...*
>
> *COVID-19, then, was both a direct consequence of the paradigm of endless growth, and the pin that burst that bubble of growth. As such, its systemic consequences have been widely underestimated—largely because the true contours of that paradigm are not widely understood.*

He thus recognizes that the "reverberations of the COVID-19 pandemic will be far-reaching and difficult to anticipate." As a corrective step to offset that difficulty, he calls on systems thinking:

> *But a number of systems frameworks offer ways of looking at the problem that can help us better conceptualize what has happened, how it could unfold, and what role we can play.*

Obviously the above statement echoes my intention in composing this Coda. To grasp Ahmed's detailed and closely-reasoned treatment of how systemic analysis reveals a potential "Gateway to the Next World," you are strongly urged to retrieve and consult the original article in *Yes! Magazine*. It is a rewarding read for anyone who is dedicated enough to have read this far in the current book!

* * *

The articles summarized above constitute a relatively small proportion of the ones I have turned up so far, yet they are sufficient to illustrate the usefulness of systems thinking in conceptualizing crises ranging from a deadly pandemic to the developing threat of fascism in a country that prides itself on being a democracy. I will lay out briefly what I consider to be the major points about systems thinking illustrated in these three articles.

In the first article Matthias Horx clearly demonstrated the CDST principles regarding: *systems far from equilibrium; turbulence; bifurcation;* and *attractors.* COVID-19 posed an immense challenge to a socioeconomic system that had evolved in a state of blindness to the possibility of such a challenge, was unbalanced in critical ways (far from equilibrium), and consequently lacked the means to cope with it effectively. This forced major aspects of the system to scramble desperately to adapt to the challenge (turbulence), attempting novel responses unavailable to the old system. Thus, bifurcation.

Horx employs a technique of creative imagination to posit a new system characterized by qualities that the previous system had failed to develop adequately. In CDST terms the bifurcation had led from one attractor to another, more evolved, attractor.

In the second article summarized above, Patrick Cook-Deegan does a thorough job of documenting a range of COVID-19's actual impacts on the functioning of everyone involved in schools in the US, and then extrapolating from those impacts to predict major changes in how education might/could be practiced after the pandemic subsides. Although he does not explicitly link those predicted changes to aspects of systems thinking, it is clear that post-chaotic bifurcation, self-organization around a new attractor, and consequent evolution of the system of education, are implicitly involved.

From my standpoint as a would-be contributor to the evolution of education, this article points to an exciting prospect. COVID -19 has disrupted the seemingly entrenched institution of public education in the US; a tipping point has been reached; the breath-taking prospect of guiding the institution in a more desirable direction (extensively discussed in this book) is therefore consequently and unexpectedly within our grasp.

Now we come to the third article, by Nafeez Ahmed, who follows an arc similar to the one posed above, but over a vastly wider territory. That is, he is addressing "the paradigm of endless growth" that underlies our entire socio-economic system, of which the educational system has been a dedicated servant.

To repeat one of his most crucial statements:

> COVID-19, then, was both a direct consequence of the paradigm of endless growth, and the pin that burst that bubble of growth. As such, its systemic consequences have been widely underestimated—largely because the true contours of that paradigm are not widely understood.

Now, put that alongside one of my introductory statements in Chapter 1:

> The framework I am about to offer is the best one I know of to advance understanding, because it is applicable to systems at every scale—including the socio-political system, the educational system, and the human nervous system. (p. 11)

That framework is of course the theory of systems, which provides the means of understanding "paradigms and how they shift." Ahmed's article is about exactly that: understanding what he terms "the true contours" of "the paradigm of endless growth." While acknowledging that "reverberations of the COVID-19 pandemic will be far-reaching and difficult to anticipate," he points out that "a number of systems frameworks…can help us better conceptualize what has happened, how it could unfold," and essentially how we can put that understanding to work in avoiding ultimate catastrophic collapse by guiding the system to a new, evolved level.

His article applies that principle to a number of systemic components of the "endless growth" paradigm, in an approach characterized as:

> …a radical reshaping of our frame of orientation, from endless material growth…to a new life-oriented system

explicitly designed for the protection and flourishing of the human species and all living beings.

To conclude this discussion of challenges precipitated by COVID -19 and the potentially creative responses to those challenges, I will describe a personal experience my wife and I recently enjoyed. First, some background: our younger son has decades of experience as a stage actor, and like most members of that profession had abruptly found himself out of work. Live theater has simply ceased, and it is impossible to predict when it might resume.

Chase Bringardner, Professor and Chair of the Auburn University Department of Theatre—in an interview conducted in May 2020 and published that month in the *Alabama Newscenter*—said about the COVID-19 challenge, "I imagine the arts will also continue to grapple with the use and presence of technology in our lives interrogating many of the very platforms—like Zoom—that have become so much a part of our daily lives."

Our son had the good fortune to join the cast of an experimental San Francisco Shakespeare Company's production of *King Lear*. The cast began rehearsing for this show about the time of Bringardner's prophetic interview, and the show premiered two days before I wrote this account. This production may well stand as the most ambitious and complex use thus far of the technology referred to in the interview.

It was exhilarating to watch this pioneering theatrical venture— pioneeering in that it was a complete and fully-staged ensemble performance of a classical play, broadcast live to an at-home audience, in which all members of the cast performed solo from *their* respective homes. Here's how it was done: The actors simultaneously performed in front of home-constructed green screens—able to hear each other but not able to view each other—while transmitting their performances to a central technical studio where they were instantaneously combined, synchronized, provided with pre-recorded background scenery and sound effects, and broadcast to a home audience.

Please take some time to appreciate the ingenuity, historic novelty, and sheer difficulty of presenting a major theatrical production in this way.

It was a stunning artistic and technical achievement, putting professional actors back to work and dazzling at least the audience of two proud parents attending the performance safely isolated in our home. What an impressive example of true intelligence and resilience in response to a public health challenge unprecedented in our time.

Final Notes

At this point, there is no telling when or how either of the crises described at the beginning of the Coda will be resolved. So I must conclude this book, in a sense, without finishing it.

I began writing it with the hope that it would have a positive effect on the system known as education—that it would advance understanding of human development in general by viewing that subject through a conceptual lens that takes into account its vast complexity, while also modeling the potential for its sudden positive change.

With the advent of COVID-19 that original intent has abruptly been infused with higher intensity and greater scope. My hope at this point is that the book will equip and empower its readers with cognitive tools needed for joining—with optimal effect—in the enormous project before us all: guiding any and all major systems that affect us directly, especially the system of education, to *self-organize at a more adaptive level* successfully, happily—and soon.

References

Ahmed, N. (2020). The Light at the End. *YES! Magazine*, Summer 2020.

Bote, J. (2020). *USA Today*, June 29, 2020. https://www.usatoday.com/story/news/nation/2020/07/29/umbrella-man-who-broke-windows-floyd-protests-white-supremacist/5535596002/

Cook-Deegan, P. (2020) Can the Lockdown Push Schools in a Positive Direction? *Greater Good Magazine*. https://greatergood.berkeley.edu/article/item/can_the_lockdown_push_schools_in_a_positive_direction

Horx, M. The Post-Corona World. https://onlineshop.zukunftsinstitut.de/shop/die-welt-nach-corona/

Appendix A

Discussion of the Initial Five Phases Diagrammed in the Chapter 7 Table Optimal Human Learning and Development

Phase One

The very first stage of intelligence to emerge in the life of a human, labeled "Sensory-Motor Integration, typically spans the first two years of life. Successfully negotiating this stage results, according to Erikson (1963), in developing hope and trust (which contemporary psychologists would term "secure attachment"). Steiner (1996), as we see at the far left of his row in the Framework table, proposed that the successful outcome of this fundamental stage would be an orientation toward—or readiness for—goodness. Maslow (1971) identified the basic needs that must be fulfilled at this stage as the needs for safety and survival. Piaget (1952), with more of an emphasis on cognitive development, named this the "sensorimotor stage", and this corresponds to development of the brain stem and cerebellum (called by MacLean [1990] the "reptilian brain" as seen at the far left of the row labeled Brain Stage.)

Glancing at the lower half of the Framework we see that the task set by nature at this initial stage of development is to "bond with and attach to a consistent, nurturing care provider." In the following row is found the optimal development strategy for this stage: to "provide a caring and supportive environment that maximizes love and safe limits while minimizing harmful stressors." By following this strategy we provide the setting needed by the growing child to negotiate this most basic Level of Emergent Intelligence—i.e., the integration of sensory perceptions and motor skills while instilling in the child a basis for trust and hope (articulated by Erikson), a fundamental sense of safety-in-the-world (the most basic need described by Maslow), and a tendency toward goodness (the ideal quality noted by Steiner [1996] established at this stage).

Further insight into the vital importance of bonding and attachment in successfully completing Phase One is provided by psychological theorist Daniel Goleman (2006), who described a "parent-child loop" as not only the chief way for a young child to receive basic training in how to conduct relationships, but also as a basis for shaping intellectual development in later stages: "The intuitive emotional lessons from the wordless proto-conversations of the first year of life build the mental scaffolding for actual conversations at age two" (p. 164).

Phase Two

As children approach the end of their second year of life, their neural and somatic development (assuming conditions have been favorable) has prepared them for the next phase of their life, which will last until about age six or seven.

Consulting the Framework we see this second brain development stage, designated by MacLean the "paleomammalian" stage (seen in the next cell over in the Brain Stage row), called by Piaget the "pre-operational stage," and so on down the top half of this column of the Table (Erikson, Maslow, etc.).

In the lower half of the Table we see, looking at the first row, that the intelligence level that emerges during Phase Two (again assuming all goes well—which will henceforth remain our default assumption) emphasizes symbolic representations, named in this cell as "mental images, pictures, words, language and stories."

Just as Phase One has corresponding linkages among the various rows in the Table's upper half, so does this Symbolic-Representational Phase Two, as it a) corresponds connectively with the Brain Stage labeled Paleo-mammalian (limbic system) maturation, b) serves Level Two of Nature's Plan by enabling the child to *overcome obstacles to development*, c) can be bolstered via the Guiding Strategies *movement, play, and imagination* that build on the sensory-motor integration accomplished in Phase One.

The abilities to play and imagine freely are perhaps the major determinants of how the structures of the forebrain develop—and to a large extent, how creative, integrative and satisfying a life might be. This would be in keeping with MacLean's Triune Brain theory: that individual brain development recapitulates the evolutionary journey, and that therefore each additional layer develops from the previous structure.

This principle can be seen in the progression from Phase One, which focuses on development of the brain stem, to Phase Two with its focus on developing the limbic system above and around the brain stem. Interactive with that brain stage progression, the cognitive and emotional capabilities also progress as we have just seen. *Each phase is built on the ones preceding it;* that is the scheme all the way up the ladder of development.

Once that principle is grasped, it will be much easier to comprehend the entire sweep of development depicted in the combination of illustrations above. In the previous two pages we have traversed some ways of flexibly using the combination to work our way upward on the ladder of human development by encompassing a complex range of lenses through which to view each step.

Again, regarding these first two phases of development in childhood, it should be of help for teachers and parents to know how critical it is that children build up structures of knowledge about their environment through *all* of their senses. In doing so, children form an extensive experience base that will in time come to undergird the culmination of Phase Two: the triumph of understanding the sensory world.

Clearly, the optimal way by which a child can gain this understanding is through close contact and interaction with the natural, social and cultural realms. The importance of attaining optimal sensory development in this manner cannot be overemphasized.

It is our responsibility as teachers and care providers to craft the learning environment so that it provides stimulation and support for young brains eager to learn, while simultaneously canceling out stressors that sabotage the learning brain. In general, the richer the environment and the more interactions with it, the more connections form and the stronger they become.

The basic requirements for enriched learning are quite simple: a nurturing environment, the absence of threat, and the presence of novelty, challenge, and feedback. When these conditions prevail, the child's playful engagement with her world is supported. The importance of play in this second stage of development cannot be emphasized enough. This includes both inner play in the form of fantasy and interactive play with both peers and care providers.

In an atmosphere supporting play—and most importantly, open-ended play where the child controls the type and pace of activity, with adults functioning, only when necessary, as guides-on-the-sides—children are encouraged to explore the world with all their senses, and allowed social interaction for most activities. Thus they become steeped in multiple systems of feedback—which is essential to learning.

Teachers and parents: despite ongoing pressures to narrow, reduce and even eliminate play from childhood, know that by putting forth your best efforts into play-time you are ensuring the best possible outcome. This also applies, by the way, to teens; for example, in surveys asking high school students to rate those factors most important to their learning, the vast majority make mention of teachers with a sense of humor and/or playful spirit.

Also essential to learning—and related to play—is movement. Movement awakens and activates many of our mental capacities (not only at this young stage of development, but throughout the life span), and anchors new information and experience into our neural networks. In short, movement is vital to all the processes by which we embody and express our learning, our understanding, and ourselves.

Generally speaking there are two kinds of movement that facilitate brain development and overall wellbeing. For simplicity's sake we can think of one kind as *ingrained* and the other as *novel*. Ingrained movements are those we have previously learned (walking, running, swimming, bicycling and so forth). Novel movements are those we are currently learning, and which therefore serve to increase neural capabilities. Examples of these include learning how to dance, play a musical instrument, practice a martial art such as aikido, or engage in therapeutic movements such as those taught by Feldenkrais practitioners. Both kinds of movement are exceedingly important.

John Ratey, M.D., a physician with an abiding interest in how exercise improves learning in children and whose SPARK program has been adopted by dozens of schools, details measurable improvements in areas of learning including perceptual skills, intelligence quotient, achievement, verbal tests, math tests, developmental level, and academic readiness.

As a potent counterpoint to the aggressive push to increase the amount of classroom time focused exclusively on academics, Ratey (2002) points out that, "Studies show that a reduction of 240 minutes per week of academic class time, *replaced with increased time for PE*, leads to higher math scores." More broadly speaking, he puts a fine point on the role of movement, saying that, "Our 'higher' brain functions have evolved from movement and still depend on it" (p. 148).

Carla Hannaford (1995) has stated, "The more closely I consider the elaborate interplay of brain and body, the more clearly one compelling theme emerges: Movement awakens and activates our mental capacities" (p. 95). To this she added:

> As infants, children teens—and, yes, we adults— learn to execute new movements, the brain benefits in myriad ways, providing the neural foundation for executing higher and more complex thinking and problem solving. This is especially the case when the new movements involve exquisite balance, coordination, postural grace. Think: dance, tai chi, gymnastics, athletics, playing a musical instrument. (p. 96)

In summary, providing ample opportunity for movement of all kind does two things for development and overall well-being—one fairly immediate and the other more long-term:

- Active movement readies the high brain for new learning;

- Stimulating movement integrates the brain bottom-to-top, side-to-side, and back-to-front, thus facilitating the growth of neural patterns that builds the brain's capacity.

The importance of downtime must also be emphasized (and again this requirement applies not only to the life stage currently under discussion, but throughout the life span). Downtime refers to the chronobiologic need to withdraw attention from the external environment in order to assimilate what has been learned there. It must be part of the mix that characterizes an ideal learning environment.

My experience with school curricula and classroom practice in the US—first as a student and later an observer—is that the importance of downtime is largely unrecognized, as is the importance of balance and integration. These unfortunate omissions have led to scheduling practices that are antithetical to the health and well-being of the embodied brains of students.

I have also noticed that most curricular content and classroom practice in the US is geared almost exclusively toward left hemisphere processing; processes involving the right hemisphere are largely ignored. This is another instance of failing to recognize the vital importance of balance and integration—in this case between cortical hemispheres.

To reiterate, a major purpose of this book is to influence the prevailing paradigm guiding education in this country—to replace corrosive goals, environments and practices with guiding principles and activities that recognize and support healthy brain/mind development. Toward that end, in this concluding Appendix, I describe a range of such guiding principles—pointing out ways in which education and parenting need to be tailored so that they may guide and support growth of embodied brain and mind that is congruent with optimal development.

A major purpose of this book is to influence the prevailing paradigm guiding education in this country to replace corrosive practices with guiding principles and activities that recognize and support healthy brain/ mind development.

Phase Three

Having successfully incorporated Symbolic-Representational intelligence into their expanding repertoire of tools for living, children in the third phase of development are firmly situated in elementary schools and ready to welcome in earnest the next level of emergent intelligence: the Social-Emotional (next column over in the lower half of the Framework). By the age of six or seven their brains' limbic systems are fully on-line (thanks to Phase Two), and the posterior parts of their cerebral cortices are in the spotlight for full development. As they grow toward puberty, the natural need to acquire emotional fluency, sharpen relational skills, and develop their imaginations becomes the major focus of their young lives. This is indicated in the third column of "Nature's Plan."

During this phase of development the scope of their lives has expanded beyond the boundaries of home, and consequently they are now exposed to a much broader range of models by which to be influenced. Ideally, the set of models available to them closely matches their expanded capacities. These capacities include creative imagination and an impulse towards collaboration and cooperation that will require honing what Daniel Goleman (2006) has termed "social intelligence."

Societal strategies for guiding and supporting young people at this level of development (as indicated in the "Guiding Strategies" row) must, at a minimum, help them recognize and learn about increasingly complex emotions and navigate increasingly complex relationships. They need to learn, for instance, how their emotions affect their perceptions, their thought processes, and even how coherently their bodies function. They need to know that they can self-modulate their moods, and be taught some basic methods for doing so. It would help them immensely to begin learning some of the attentional skills discussed in Chapter 5.

As for growing into emerging intelligence about relationships at this level, play provides an excellent platform, especially play that involves taking different roles. And play that requires vigorous movement is doubly effective in that it increases oxygen-rich blood flow to the brain (improving mental clarity), elevates serotonin for balanced moods, and reduces stress and depression.

The environment within which this learning occurs must be carefully considered, with regard to issues raised in *Anatomy of Embodied Education* such as circadian and ultradian cycles, freedom from threat, and the vital importance of relationship. Of course, these issues pertain to every level of development. I emphasize them here because their relevance reaches a peak during Phase Three.

Diligent consideration of developmental factors—especially those beginning with Phase Two—not only enhances learning but also positively affects the overall well-being of learners. Consequently, and very importantly, it can be expected to reduce discipline problems.

Phase Four

As humans complete Phase Three, emerge from puberty and commence their Phase Four teen and early adult years, the perceptual and motor parts of their posterior cortex are well-developed and humming along. The frontal lobes of the cerebral cortex become the next focus for brain development.

Let's look next at the level of Emergent Intelligence associated with Phase Four, named "Concrete-Creative." The table shows us that important activities at this stage involve multi-sensory manipulation, experimentation, building, and creating.

After successfully laying the necessary foundation in the first three phases of development, these young people now face the tasks set by Nature's Plan that enable them to both differentiate and integrate multiple intelligences. The endless process of self-discovery takes a huge leap at this level as creative outlets are experienced and the beginnings of mastery are made possible.

Those of us responsible for creating adequate settings and processes for learning opportunities may find it particularly challenging in the current educational climate to meet these challenges productively. While well-intended, the almost exclusive emphasis on intellectual achievement at the secondary school level along with a simultaneous neglect of body and heart intelligences, has resulted in a serious impediment to balanced development.

Piaget (who, as indicated in the upper half of the Framework, termed this stage "Formal Operations"—distinct from "Concrete Operations" in the previous stage) recognized the need at this stage for an education that stresses equal measures of what we might call "hands, heart, and head." By this we mean that mental operations reaching fruition at this level are deeply rooted in what the learner has gained during the first three phases through direct sensory development, plus skills based on movement, imagination, emotions, and relationship.

Of course learning involving the "head" part of the triad will not have been neglected at the earlier stages, but ideally not overemphasized either. Neither will the "head"—intellectual knowledge and information-processing—be ignored at Phase Four. All three need to be addressed, albeit in a balanced context, with equivalent emphasis on "hands and heart" as a precursor for what it is to be focal at Phase Five.

Suggested Guiding Strategies for this phase include full use of both arts and sciences curricula, building in opportunities for creative exploration and expression. This is so that Nature's Plan at this stage of development—the discovery and expression of creativity and of an expanding range of multi-faceted intelligence—may be fulfilled.

Even learning in the "head" category needs to be more nuanced than it usually is in the US, where school curricula and classroom practices in these learning environments are geared largely to left hemisphere activities. My impression is that school systems in the US too often forget (if they were ever aware) that both hemispheres of the brains of their students need to be cultivated.

Regarding the subject of embodied learning as a vital focus of Phase Four development, Tim Burns and I in our earlier book note that disciplined effort devoted to embodied learning often results in evocative and even charismatic qualities, a sense of ease and gracefulness in such endeavors as performance art, music and athletics. These high-level accomplishments must not be sacrificed, for to do so invites degradation of our entire culture.

Phase Five

Learners who have mastered the lessons and challenges of Piaget's cognitive stage known as Formal Operations, with their cerebral cortices integrated and well-balanced, are now poised to enter the cognitive stage called Post-formal Operations. In their late teens they are nearing completion of secondary education and preparing for young adulthood. Brain development at this stage consists primarily of refining connections among all three of the brains discussed in Chapters 11 and 12 of *Anatomy of Embodied Education*, with special attention to bringing heart energy into mental processes.

The intelligence emerging at this level, as indicated on the lower half of the Framework, features the abstract and conceptual. Learners are prepared to enter the heady realm of constructing and testing models about themselves and the world in which they are embedded. Not only will they learn logic—how to hypothesize, reason and analyze—they will also learn metacognitive skills such as suspending assumptions and thinking beyond conventional limits.

These new abilities will enable them to perceive and gauge what is relevant and significant in their young lives as nature's plan continues to unfurl. They are learning to think confidently and collegially, and to find depths of meaning that will serve them throughout their lifetimes.

Helping to guide learners at this level consists primarily of modeling the abilities that this cognitive level requires them to attain. Joseph Pearce calls this the model imperative.

Learners will benefit from engaging in tasks that involve all the cognitive skills named above. Furthermore, in order to give the best possible rehearsal for the life to follow, the tasks should be themselves genuine, authentic—projects that are relevant to the real world.

Below is a summary list of best practices recommendations related to authentic and real world involvement for teens, practices that are both time-honored and ubiquitous across societies. According to research on protective factors related to both academic success and adolescent wellbeing (Benard, 1991, 2004; Burns, 1994) the following practices/

environments are known to play a powerful role in facilitating the resiliency response as well:

> mentorships; peer-to-peer teaching/coaching;
> service learning;
> cooperative learning;
> adventure learning;
> sufficient downtime.

At this stage of development and beyond, the frontal lobes of the cerebral cortex continue to create new networks, then link these networks to others in the cerebrum, the heart-brain, and the gut-brain. In other words, more neural complexity is being constantly created, bringing with it a variety of coping abilities requisite to the challenges learners are bound to face.

References

Benard, B. (1991). *Fostering Resiliency in Kids: Protective Factors in the Family, School and Community*. Western Regional Center for Drug-free Schools and Communities.

Burns, E. (1994). *From Risk to Resilience*. Marco Polo Publishers.

Goleman, D. (2006). *Social Intelligence: The New Science of Human Relationships*. Bantam.

MacLean, P. (1990). *The Triune Brain in Evolution: Role in Paleocerebral Functions*. Plenum Press.

Maslow, A. (1971). *The Farther Reaches of Human Nature*. Viking Press.

Piaget, J. (1952). *The Origins of intelligence in Children*. International Universities Press.

Steiner. R. (1996). *The Education of the Child in the Light of Spiritual Science*. Anthroposophic Press.

Appendix B

Reflections on an Above Average Public School Education
by Marika L. Foltz

This appendix begins with one reader's essay on her responses to the book as applied to her own history as a student in public schools in Northern California. In it she describes concrete examples of atmospheres and expectations she encountered in schools, and draws distinctions between those that impeded her and those she found helpful.

She is now the mother of a daughter, Sandra, who is going through early stages of that journey (currently in lock-down because of COVID-19).

Marika has told me in person that the material covered in the book increases her confidence about the choices she and her husband have made about the unconventional course of their daughter's education. Reading the book apparently deepened her understanding of the issues that underlie how unsatisfactory her own school experiences were compared to those happily in store for her daughter.

I am glad for that outcome, and hope that it—or its equivalent—typifies the response of most who read this. Even more, I hope that this book contributes to a serious reconsideration of how the system of education is conceived and manifested in the US and other developed countries in the Western World.

Marika affirms and joins that hope in the concluding extension of her essay, in which she widens its focus beyond her own experiences as a student in a conventional school system. In those final pages she recounts more generally her reflections about the disruptive impact of the pandemic on western education as it has existed until now.

Looking back on my public school education, I know I should feel fortunate. In many parts of the world, even today, the right to become educated is by no means a given. I was brought up in a small, middle class city in Northern California in the 1970s and 1980s. The school environment was carefully nurtured and most if not all of the administrators and teachers were conscientious in their endeavors. It was also a sheltered environment, far removed from the stressors of inner-city life. It was even a school district with a high rating based on national test results, and many people wanted to move to our town because of its schools. However, while acknowledging with gratitude the good fortune I had in receiving what could be called an above average education, I also look back on it with a critical eye. Every person is different, and I am sure there must be people who thrived in a similar academic environment. However, *for me*, what I can honestly and fairly say is that despite the best intentions, there was a lot of room for improvement.

In the early years, in elementary school (K-5), I can pinpoint a few key reasons why the education I received was not ideal for me. The first and perhaps the most crucial one was that I may have started Kindergarten too soon. I had an October birthday, so instead of waiting until I was almost six, I started school in September when I was almost five—which means I was still four. Of course there are four-year-olds and four-year-olds. I was not the most outgoing, confident, or mature four-year-old. In hindsight, I am convinced that I was at a gross disadvantage due to my being one of the youngest, and more importantly, my being too young for what was expected of me both academically and socially.

Academically, starting Kindergarten at age four may not appear to have held me back—I got good grades, I went on to attend a good university, I got a degree, and I have had no problems finding work. Nevertheless, I have always been an especially slow reader. Reading was always a tedious task for me, and while some kids could do their homework in no time flat, it took me ages. There may well have been other factors involved in my becoming a slow reader, but I am convinced that my lack of enthusiasm for reading when I was younger and my anguish around assignments involving long reading passages dates back to my being expected to learn to read when I was still five (in first grade) and my being developmentally unprepared for it. If I had started school one or even two years later, how would I be different today? Would I be a more avid reader? Would I be a Pulitzer Prize winner? Maybe not, but you just never know.

Even socially, starting school so young may have been a hindrance to me. I recall more unhappiness in school related to social situations than to academic ones. Could my relative immaturity (socially and developmentally speaking) have impaired my ability to make friends easily, to have a positive sense of self-esteem, or to be thick-skinned when insults were hurled at me, for example? Would I have thrived more if I had been older when faced with the same social situations and challenges? Again, I do not know the answers to these questions, but I cannot help but wonder.

What I can claim about my above average public school experience is that for me, it marked the beginning of being immersed in a competitive environment, the beginning of a feeling of inadequacy, and the beginning of stress.

By far the worst subject of all, in all of my years in school, was Physical Education (P.E.). I felt comparatively ungifted in strength, agility, balance, limberness, speed, stamina, aim, and gracefulness, so P.E. was not my forte, to say the least. But that did not excuse the teachers' affinity for turning P.E. class into a popularity contest—or worse yet, a game show. They would invariably choose team captains who would in turn choose their team members—one...by...one. This excruciating selection process, with a highly predictable outcome, could not possibly help me or other non-athletes to develop a sense of comradery, sportsmanship, self-acceptance, or self-esteem—or, need I say, a love for sports. And I fail to see how feeling miserably inadequate helps one become better at sports, or at anything else for that matter. Only one year—one semester to be precise—did my P.E. teacher decide to give grades based on effort and attitude. That was the one semester I ever got an *A* in P.E., much to my surprise and delight.

Another P.E. highlight, now in middle school (6th – 8th grades), I can clearly remember the national testing procedure whereby we had to run for two miles in scorching heat, not having trained for it adequately. Being the next-to-last one to reach the finish line was not an issue for me. I was used to that. My biggest concern was whether or not I would drop dead before reaching the end of the race course. My head puffed up like a swollen tomato, I arrived panting, wheezing, and very nearly collapsed after crossing the finishing line. This is what comes from treating every student the same, turning us all into a mere statistic. We are *not* all the same. I am extremely good at foreign languages, and remarkably bad at sports. Why do we all have to be taught as if we were robots, expected to perform exactly the same as each other, at the same time, and in the same way? Don't grades and testing only undermine self-esteem by jabbing us when we are down? In general terms, we all know what our weaker subjects are. Couldn't more teachers be like that one lone P.E. teacher who took effort and attitude into account? I am trying to imagine a scenario in a math class where team leaders are chosen from among the best students in the class, and they in turn choose their classmates—one…by…one—to join their team to solve complex math problems. Some people just have an incredibly difficult time with math. How would they feel being left until last *every* time in the selection process? Of course in a math class this scenario sounds absurd, but in P.E. it was the norm. If the education system would accept and even *value* the fact that we are not all the same, I believe we would be much better off as individuals and as a society.

Middle school had its better and worse moments. The school play in 8th grade, learning Spanish, and making new friends were among the highlights. Social stressors and P.E. still loom large in my memory, but I was academically happier starting in 7th grade, when more diverse subjects were introduced and more challenging classes were offered to students who could handle them. Finally, there was some acknowledgment that we are not all the same. Of course tracking has its risks, and it is undeniably dangerous to place someone in a "remedial" class too soon, as they may never get off that track. But for me at least, being in a more challenging environment in certain classes marked the difference between apathy and enthusiasm. And yes, I definitely would have appreciated being placed in a "remedial" or "basic" P.E. class that respected my aptitudes, and being treated with respect despite this placement.

I have far more memories from high school than from previous years. Two things I can say with all honesty: that the most useful subject I studied in high school was typing, and that the best thing about high school was graduating. I do not mean this facetiously. I am completely serious. Typing *was* the single most useful skill I learned. It has helped me in every other subject I ever enrolled in and in many outside activities as well. And the teacher did not even need to be that good. You just needed to show up and dedicate time to it. And when I say that graduation was the best thing about high school, that is just to illustrate, truthfully, *how* burnt out I was by the end of those four years. Burnt out at age seventeen!

I studied hard in high school. So hard that I had no free time to relax. So hard that I seldom slept more than six or six and a half hours on weeknights. This was mainly because of the early mornings and all of the homework I had, which I generally felt was busy work—otherwise known as a waste of time and energy. (It is now believed that teenagers should get 11 hours of sleep per night, just like small children. So by that standard, I was *extremely* sleep deprived). The exams were largely based on rote memory, and multiple choice tests were common. While social stressors still abounded, the competitive nature of classes, concern about my G.P.A. and the S.A.T., getting into college, and other academic concerns were even a bigger source of stress for me. This cocktail of stress and pressure and sleep-deprivation led me to make some unwise and even unethical decisions that today I regret, such as cheating on some tests to get from a B+ to an A- and thus improve my G.P.A., and choosing to take less challenging classes to avoid the exhaustion involved in taking more advanced ones (and get better grades, of course). For example, I never studied Physics because it was optional and in my senior year I felt I just could not handle any more highly demanding classes. I feared it would push me over the edge. And it might have. But still, I regret not having studied Physics, as I regret taking Basic Algebra 2 rather than Advanced Algebra 2 as advised by my math teachers. Poor decisions that

I believe were fueled by the strong feeling of burn-out that was surging within me, coupled with sleep deprivation and stress.

Another point I have reflected on is the teachers I had over the years. Having worked as a teacher myself for many years, I know how challenging it can be to meet students' needs, motivate them, assess their work, and comply with bureaucratic requirements. I also know how, especially in the US, teachers are notoriously underpaid and in many ways undervalued. Contrary to popular belief, teachers are extremely important, much more so than they are often given credit for. They can make the difference between a student loving a subject or hating it, pursuing it or avoiding it in the future. Some of my teachers were very good, even excellent, some I confess were very bad, but most were somewhere in between.

In trying to figure out what qualities in my teachers made me come away from their classes with appreciation, apathy, or disgust, I have noted a pattern that shifted around the 7th-8th grade. In the early and middle years (K-7), the teachers I recall with most fondness were the ones that could be described as "warm." I also have positive memories of those teachers who used humor in the classroom, though less so than the truly warm-hearted teachers. The rest I can just describe as neutral, or "nothing to write home about." Starting in the 8th grade, the teachers I valued the most were those who genuinely challenged us to think, to create, to push ourselves just a little harder. They were hard-working and dedicated, and very knowledgeable about their subjects. They expected a lot of us and treated us almost like college students. By approaching their subject with the utmost respect and holding high expectations for the students, we were motivated to work harder and learn more, while appreciating the topic more than we would have otherwise.

Having heard some of my qualms about the education I received, you may not be surprised to hear that, now that my husband and I are parents of a school-aged child, we have chosen to send our daughter to a Waldorf school. Many frown on sending their children to private schools, as they say it is not being supportive of the public school system. Of course I would not have public schools disappear, just evolve. If they were doing things in a more holistic, creative, artistic, supportive, and less competitive way, I would be more inclined to send my child there. And there are communities that have managed to get charter schools (publicly funded) in their towns which follow a Waldorf methodology. Kudos to them. What my husband and I really appreciate about the Waldorf methodology is that it is based on a deep respect for the emerging individual. Each child is viewed as a spiritual being who brings something unique to the planet, and that uniqueness has to be respected, nurtured, and drawn out delicately over time. This is considered tantamount for making a better world. There is great emphasis on observing each child to understand who they are and what they need in order to grow as a person (not just academically). Music, art, handwork, contact with and respect for nature, creative thinking, and collaboration are all key elements of this education system that we appreciate above and beyond good test scores and knowing a lot of facts that the students will have forgotten by the time they are my age anyway (I know I have). So far we are exuberantly pleased with our daughter's education—and even more importantly, *she* is.

Even though I was in grade school over 30 years ago, and many things have changed since then, I wonder how many of my complaints have been addressed systemically since then. I am certain that many teachers do their best to innovate, meet students' needs, and attempt to overcome ingrained, outdated methods and conceptions. However, given all of the responsibilities placed upon them, one can hardly expect them to take on Goliath in their spare time. I hope most earnestly that the education system, whether public or private, can and will change to support the individuality of each student, supplying them with a healthful environment—not just physical and mental but also emotional and spiritual—and replacing competition, rote memory, busy work, animosity, and stress with collaboration, team-work, creativity, motivation, and harmony. What is healthy for our children is healthy for our future—as a society, a species, and a planet. And surely *that* is a goal worth working toward.

Further Reflections

Since I finished writing this short piece about my own "above-average public school education," Covid-19 and all of its permutations have hit the world like a bomb. I find myself reflecting on what I wrote then, and wondering how this new situation will affect education as we come to terms with this new reality.

The Covid-19 outbreak has obliged most of us to reevaluate our priorities, adapt to new realities, and be flexible even in light of the most changeable and unsettling circumstances. These abilities to adapt and realign to the environment underscore the need to prepare students for life. How can we tackle life when it gets tough, and even tougher, unless we have an adequate education and upbringing?

A deeper kind of learning that prepares us for life involves the attainment of higher degrees of self-realization, not just academic but also moral and spiritual. This kind of mastery can give us hope for a brighter future for our planet. Gandhi's famous quote reminds us to "be the change you wish to see in the world." And education and upbringing are probably the two most important ways we have of preparing young people so that they may become ever happier, healthier, and more compassionate. In effect, we need intelligent, engaged adults who can make wise decisions in the interest of our planet and our future. Stressed out, anxious, apathetic, alienated, embittered, competitive, aggressive youth are just not going to bring about the future our world needs.

With this in mind, I have been thinking about how we might envision the re-emergence of education after the Covid-19 disruption has passed. Like the phoenix rising from the flames, the hope is that—following a brand new attractor—our mainstream education system might emerge with vital changes that will foster greater well-being and genuine intelligence based on adaptability, resilience, and creativity. Young people are (let's not forget) tomorrow's leaders. Clearly, it is crucial to embark on such a transformation with our eyes open and make sure we are aware of what elements need to go into such a novel reform of our antiquated education system.

Covid-19 has greatly affected so many sectors of society, and education has not been spared. As I write this, the vast majority of schools (K-12) as well as community colleges and universities around the US are being run with the aid of online platforms and programs for conference calling in lieu of in-person education, or some hybrid model. Of course we do not know how long this will last. It might not last more than a few months, or on the contrary, it may become the "new normal."

It seems that now is the perfect time to envision what we would like education to look like in the future and begin to move in that direction. Looking forward, I am compelled to ask myself how we can create a meaningful, motivating, invigorating new education system with a screen placed between the student and the educator, between the students themselves, and between students and the world at large?

Indeed, how can we human beings in general thrive without in-person contact with others? I feel that we cannot. Children need to be with other children in order to master the social and emotional skills so necessary for human development, and they need to have a nurturing, motivating role model to inspire them to learn. I seriously doubt that this can happen effectively over a screen. It is true that as students get older, they become more and more able to deal with remote learning and its challenges. But how will they learn to handle life's complicated interpersonal encounters and conflicts if they engage with other students and teachers only on conference calls, where they can just turn off the sound or the video and avoid involvement altogether with the click of a key on a keyboard? For me, the answer is clear. They will not.

I postulate that in-person, human contact between people—who are mammals and social beings above all else—is key to optimal human development and learning. That may be too simplistic an analysis, and I am sure there will be many who will find some clear benefits to online education. However, I do find it hard to support an online learning model as a "new and improved" one if the goal is as far-reaching as human transformation and transcendence. I stand, like the rest of humanity, expectantly facing the future, optimistic (to a point) about the potential for some kind of positive adaptation of our species in general and of our children's education in particular, even under extreme conditions. For the sake of our planet and all of us who are now alive, as well as those who will come after us, I truly hope this will be so.

—*Marika L. Foltz*

Bibliography

Abraham, F. (2015). A Beginner's Guide to the Nature and Potentialities of Dynamical and Network Theory, Part 1. *Chaos and Complexity Letters 9* (2). Nova Science Publishers.

Ahmed, N. (2020). The Light at the End. *YES! Magazine*, Summer 2020. https://www.yesmagazine.org/issue/coronavirus-community-power/2020/05/11/coronavirus-community-power-survival/

Bateson, G. (1972). *Steps to an Ecology of Mind*. Ballantine Books.

Bateson, G. (1979). *Mind and Nature: A Necessary Unity*. Bantam Books.

Benard, B. (1991). *Fostering Resiliency in Kids: Protective Factors in the Family, School and Community*. Portland, OR: Western Regional Center for Drug-free Schools and Communities.

Bote, J. (2020). *USA Today*, June 29, 2020. https://www.usatoday.com/story/news/nation/2020/07/29/umbrella-man-who-broke-windows-floyd-protests-white-supremacist/5535596002/

Burns, E. (1994). *From Risk to Resilience*. Marco Polo Publishers.

Burns, T. & Brown, J. (2021). *Anatomy of Embodied Education: Creating Pathways to Brain-Mind Evolution*. Psychosynthesis Press (Originally published by Inspired by Learning, 2020).

Buzsáki, G. (2006). *Rhythms of the Brain*. Oxford University Press.

Combs, A. (2010). *Consciousness Explained Better: Towards an Integral Understanding of the Multifaceted Nature of Consciousness*. Paragon House.

Cook-Deegan, P. (2020) Can the Lockdown Push Schools in a Positive Direction? *Greater Good Magazine*. https://greatergood.berkeley.edu/article/item/can_the_lockdown_push_schools_in_a_positive_direction

Deikman, A. (1973). The Meaning of Everything. in R.E. Ornstein (Ed.), *The Nature of Human Consciousness*. W.H. Freeman.

Edelman, G. (2004). *Wider Than the Sky: The Phenomenal Gift of Consciousness*. Yale University Press.

Ennis, C. (1992). Reconceptualizing Learning as a Dynamical System. *Journal of Curriculum and Supervision*. 7(2), 115-130.

Erikson, E. (1963). *Childhood and Society*. Oxford University Press.

Fehmi, L. (2003). Attention to Attention. *Applied Neurophysiology and EEG Biofeedback*, J. Kamiya. (Ed.). Future Health Inc.

Fehmi, L. and Robbins, J. (2008). *The Open-Focus Brain: Harnessing the Power of Attention to Heal Mind and Body*. Penguin Random House.

Gatto, J. T. (2000). *The Underground History of American Education: A Schoolteacher's Intimate Investigation Into the Problem of Modern Schooling*. Oxford Village Press.

Giroux, H. (2016). *"We No Longer Live in a Democracy": Henry Giroux on a United States at War With Itself*. TruthOut. https://truthout.org/articles/we-no-longer-live-in-a-democracy-henry-giroux-on-a-united-states-at-war-with-itself/

Gleick, J. (1987). *Chaos: Making a New Science*. Viking Penguin.

Goldstein, J. (1999). Emergence as a Construct: History and Issues. *Complexity and Organization*. 1(1), 49-72.

Goleman, D. (2006). *Social Intelligence: The New Science of Human Relationships*. Bantam.

Hannaford, C. (1995). *Smart Moves: Why Learning is Not All in Your Head*. Great Ocean Publishers.

Horx, M. The Post-Corona World. https://onlineshop.zukunftsinstitut.de/shop/die-welt-nach-corona/

Juarrero, A. (2020). Complex Dynamical Systems Theory. www.cognitive-edge.com.

Kelso, J. A. S. (1995). *Dynamic Patterns: The Self-Organization of Brain and Behavior*. The MIT Press.

Kelso, J. A. S. (2008). An Essay on Understanding the Mind. *Ecological Psychology*. 20(2), 180-208.

Kincheloe, J. (ed.) (2005). *Classroom Teaching: An Introduction*. Peter Lang.

Koestler, A. (1968). *The Ghost in the Machine*. The MacMillan Company.

Kohlberg, L. (1984). *The Psychology of Moral Development: The Nature and Validity of Moral Stages (Essays on Moral Development, Volume 2)*. Harper & Row.

Krippner, S. (1991). Inaugural Conference of the Society for Chaos Theory in Psychology, Saybrook institute, San Francisco.

Maslow, A. (1971). *The Farther Reaches of Human Nature*. Viking Press.

Miller, Ron. (1997). *What Are Schools For?* Holistic Education Press.

Piaget, J. (1952). *The Origins of intelligence in Children*. International Universities Press.

Pollan, M. (2018). *How to Change Your Mind*. Penguin Press

Ratey, J. J., (2008). *SPARK: The Revolutionary New Science of Exercise and the Brain*. Little, Brown and Co.

Ratey, J. J. (2002). *A User's Guide to the Brain*. Random House.

Schwartz, J. & Begley, S. (2002). *The Mind & The Brain: Neuroplasticity and the Power of Mental Force*. HarperCollins.

Siegel, D. (2017). *Mind: A Journey to the Heart of Being Human*. W. W. Norton & Co.

Siegel, D. (2011). *The Neurological Basis of Behavior, the Mind, the Brain and Human Relationships*. Climate, Mind and Behavior Symposium, Garrison Institute. https://www.youtube.com/watch?v=B7kBgaZLHaA

Siegel, D. (2010). *The Mindful Therapist*. W. W. Norton & Company.

Siegel, D. (2012). *Pocket Guide to Interpersonal Neurobiology: An Integrative Handbook of the Mind*. Norton.

Singer, W. (2009). The Brain, a Complex Self-organizing System. *European Review*. 17(2), 321-329.

Steiner. R. (1996). *The Education of the Child in the Light of Spiritual Science*. Anthroposophic Press

Thompson, E. (2014). *Waking, Dreaming, Being: Self and Consciousness in Neuroscience, Meditation, and Philosophy*. Columbia University Press.

Vishnoi, N. (2015). *Nature, Dynamical Systems and Optimization*. Off the Context Path. http://www.offconvex.org/2015/12/21/dynamical-systems-1/

von Bertalanffy, L. (1968). *General System Theory*. Braziller.

Index

A

Abraham, Frederick D., 40-41, 44

Abraham, Ralph, 27-28, 161

Anatomy of Embodied Education, 3, 13, 35, 39, 49, 58-59, 63-64, 66, 74, 75, 77, 89-91, 97, 136, 138

Assagioli, Roberto, 18

attention, 15, 19, 34, 52, 55-56, 58, 60, 65-73, 78, 85-86, 94, 101, 103, 106, 117, 119, 134, 138

attractor, 36-39, 42-45, 49-50, 54-56, 60, 107, 116, 118, 124, 147

    fixed point, 36-37

    limit cycle, 36-37

    Lorenz, 37, 45

autonomic nervous system (ANS), 39;

    sympathetic, 39;

    parasympathetic, 39

autopoietic, 59, 63

awareness, pure, 4

B

Bateson, Gregory, 76, 78-80, 110

Bauer, Roman, 62

behaviorism, 17

bifurcation, 37-40, 42-44, 48-51, 107, 110, 116, 118, 124

biofeedback training, 19, 28

    neurofeedback, 28, 37, 71, 86, 97, 163

brain, i, v, 3, 7-9, 13-16, 27-30, 34, 37, 45, 48-49, 51-53, 55, 57-67, 71, 73, 75-77, 79-87, 91, 100, 103-104, 109, 129-136;

    embodied, 3, 7-8, 15, 34, 45, 51, 53, 58, 62, 64-66, 73, 75-77, 79, 81, 87, 134, 137

    mind/brain connection, 14;

    Triune Brain theory, 77, 131, 139

Brewer, Judson, 105

Brown, Molly, v, 4-6, 17-18, 21, 27, 161, 161

Bugenthal, James, 19

Burns, Tim, 3, 8, 13, 28, 74-75, 90, 94, 137-139, 161

Business As Usual, 13

Buzsáki, György, 29, 30, 35, 50; *Rhythms of the Brain*, 29, 35, 50

C

Carhart-Harris, Robin, 104-105

chaos theory, 11, 27, 30

chaotic, 15, 36, 44-45, 47

Combs, Allan, 64

155

complexity, ii, 16, 25-30, 32, 34, 54, 83, 91, 110, 127, 139

consciousness, i, 3, 5-9, 13-14, 17-19, 27-28, 34, 42, 45, 51-52, 58-66, 71-73, 76, 78, 90, 97, 103, 107, 163;

nature of, 3, 7, 42, 76

constraints, 16, 55, 60

control parameter, 40-41, 43-44

coordination dynamics, 47-48, 50-53, 64-66, 99-100

COVID-19, vii, 10, 113, 115, 118-119, 122-127, 140

cybernetics, 76

D

default mode network, 103-106

Deikman, Arthur, 18-19, 78-80, 161

Descartes, Renée, 65, 76, 100

differentiation, 50, 85

E

Edelman, Gerald, 3

education, purpose of, 3, 7, 12, 28, 33, 74, 75, 121;

public, 3, 11-12, 33, 71, 124, 140-142, 145-147

ego, 4-5, 104-106

Einstein, Albert, 33, 89

emergent process, 14-15, 53, 62, 81, 83

energy, 14-15, 20-22, 37, 47, 53, 61-63, 73, 77, 79-84, 86, 103-104, 138, 144

flow of, 14-15, 77, 84, 86

Ennis, Catherine D., 54-56, 60

entheogens, 7, 103, 107

Erikson, Erik, 90-91, 112, 129-130

F

Fadiman, James, 18, 161

feedback, 22-26, 28, 60, 63, 82-83, 132

negative, 24, 30

positive, 24-26, 30-31

Fehmi, Les, 66-73, 86-87, 106, 161

flow, 14-15, 41, 53, 61-62, 77, 79, 81, 83-84, 86, 102, 105, 135

Floyd, George, 114

Frager, Robert, 19, 161

G

Gatto, John Taylor, 12-13

Giroux, Henry, 33

Gleick, James, 28, 44-45

Goldstein, Joseph, 41

Goleman, Daniel, 112, 130, 135, 139

H

Hannaford, Carla, 133

holarchy, 21

holon, 21, 24, 36-37

human development, 3, 7, 9, 12, 14, 28, 89, 91, 94-96, 98-101, 107-108, 127, 131, 148

human potential, 2-3, 6, 9, 17, 67, 98, 109

humanistic psychology, 1, 17, 98

I

identity, 4-5, 7, 87, 102-104, 106

information flow, 14, 15, 53, 61

integration, 48, 50, 70, 75, 84-87, 94, 97-98, 129, 131, 134

intelligence, 3, 7, 9, 12-14, 16, 29, 35, 42, 45, 52, 58, 65, 73-74, 89, 91, 94-97, 108, 112, 117, 127, 129-130, 133, 135, 137-139, 147

intention, 34, 51-52, 58, 65-66, 72-73, 89, 115, 123

J

Jaffe, Dennis, 19

James, William, 51, 64, 66, 72, 76

Juarrero, Alicia, 52

K

Kelso, J. A. Scott, 47-53, 56-57, 64-66, 71, 99-100, 112

Kincheloe, Joe Lyons, 32-33

Klimo, Jon, 27

Koestler, Arthur, 21

Kohlberg, Lawrence, 90, 112

Krippner, Stanley, ii, 19, 45, 161

L

learning, i, 1-2, 9, 11, 19, 25, 29, 33, 40, 51-52, 54-57, 60, 66, 71, 83, 102, 104, 111, 118-121, 131-139, 143, 147-148

LeShan, Lawrence, 19

linearity, 31, 32, 34

    non-linearity, 11, 16, 28-31

Lorenz, Edward, 45

LSD, 5-7, 106-107

M

MacLean, Paul, 77, 91, 129-131, 139

Macy, Joanna, 21, 161

Maslow, Abraham, 9, 48, 90, 98-100, 107-109, 112, 129-130, 139

Miller, Ron, 12-13

mind, v, vii, 3, 7, 9, 11, 13-16, 29, 34, 42, 45, 47-49, 51-53, 55, 57-59, 61-65, 67, 72-86, 90-91, 96-97, 100, 104, 108, 111-112, 134, 147

    mind/brain connection, 14,

mind/body relationship, 90-91

mindfulness, 70, 72-73, 87, 97, 102, 111

N

Naranjo, Claudio, 19

Nature's Plan, 96, 97, 101, 130, 136, 137

nested hierarchy, 21

neural synchrony, 63-64

neurofeedback, 28, 37, 71, 86, 97, 163

neuron, 29, 62

neurophilosophy, 4

neuroscience, i, 4, 8, 13, 19, 34, 52, 60, 76, 90, 95, 97, 108-109, 163

Newtonian physics, 31

No Child Left Behind (NCLB), 32, 120

O

Ornstein, Robert, 18

P

peak experience, 5-6, 98, 102, 105, 107-108

perception, 60-61, 78, 102, 106

phase, 9, 39, 43-44, 49-51, 60, 66, 70-71, 86, 101, 107-108, 130-131, 135, 137

portrait, 44, 51

shift, 39, 43

Piaget, Jean, 90-91, 112, 129-130, 137-139

Pollan, Michael, 103-106, 112

psychosynthesis, 18, 97

R

Ratey, John, 133

reductionism, 19-20, 32-33

relationship, i, v, 15, 19, 29, 34, 41, 53, 58, 62, 72, 75-76, 78, 80, 82, 98, 100, 111, 117, 123, 136-137

S

Salner, Marcia, 19, 161

Saybrook Institute, 19, 27, 161, 163

Schwartz, Jeffrey, 72-73

self, 15, 61, 87, 97, 104-106, 110

ego, 4-5, 104-106

self-actualization, i, 9, 48, 98-101, 107-111

self-organization/self-organizing, 15, 22, 24, 25, 40-41, 47-49, 59, 61-63, 81, 83-84, 110, 124

self-regulation, 22, 82, 110

sentience, 63

Shapiro, Stewart, 17, 161

Siegel, Daniel, 11, 14-16, 28, 53, 61-63, 75, 77-81, 83-87, 100, 102-104, 110, 112

Pocket Guide to Interpersonal Neurobiology, 75-79, 85

Singer, Wolf, 53

stability, 47-48, 54-55, 122

  instability, 47-48, 50, 55, 116

  metastability, 49

standardized tests, 32

Steiner, Rudolph, 90, 112, 129-130, 139

synchrony, 51, 63-64, 71, 86

systems, i, 1, 7-11, 13, 15-16, 19-22, 24-33, 35-37, 39-41, 43-45, 47-49, 52-55, 57, 59-60, 63, 70-71, 76, 79, 81-83, 96, 101, 110, 113, 115-116, 118, 122-125, 127, 132, 135, 137, 163

systems theory, 11, 13, 16, 19, 21, 27-28, 35, 44-45, 53-55, 59, 76, 81-82, 101, 113, 115, 163;

  general, 11, 19, 76, 81, 86;

  complex dynamical, 7-8, 16, 24, 27-28, 30, 35, 38. 45, 48, 54, 59, 65, 81, 101, 113, 115;

  living, 21-22, 24, 37, 39-40, 51

T

Tart, Charles, 18

Thompson, Evan, 59-64, 71

trajectory, 17, 36-38, 42-43, 50, 116, 163

transpersonal, 2, 18-19, 96-98, 107, 163; psychology, 18-19, 163

V

Van Regenmorte, Marc H.V., 32-33

vector, 42

volition, will, 72

von Bertalanffy, Ludwig, 19-20, 27

About the Author

Jim Brown, PhD, has spent his entire professional life quietly patrolling the frontiers of discovering and applying knowledge that involves consciousness and neuroscience. He recognized and seized the opportunities to become an early adopter in burgeoning fields such as biofeedback, then neurofeedback, plus theory and practices stemming from humanistic and transpersonal psychology, plus steeping himself in systems theory as a context for all his other fields of focus.

Jim holds a MS in psychology from San Francisco State University and a PhD from Saybrook Institute. His previous publications include *Language Be My Bronco: A Life in Poems* (Psychosynthesis Press, 2012). He lives in Mount Shasta CA with Molly, his life partner for over 60 years. They have two sons and two grandchildren.

The lodestar of my life—the search for the deepest understanding of life's purpose and of pathways toward fulfilling that purpose —has led me finally to writing this book after co-authoring its predecessor, Anatomy of Embodied Education. *I hope that in some way their combined contents can help humanity and the planet that is our matrix survive and thrive together.*

—Jim Brown

CPSIA information can be obtained
at www.ICGtesting.com
Printed in the USA
LVHW011242290921
699025LV00013B/816

9 780991 319619